STUFFED ANIMALS & PICKLED HEADS

STUFFED ANIMALS & PICKLED HEADS

THE CULTURE AND EVOLUTION OF NATURAL HISTORY MUSEUMS

STEPHEN T. ASMA

OXFORD
UNIVERSITY PRESS

2001

OXFORD
UNIVERSITY PRESS

Oxford New York
Athens Auckland Bangkok Bogotá Buenos Aires Calcutta
Cape Town Chennai Dar es Salaam Delhi Florence Hong Kong Istanbul
Karachi Kuala Lumpur Madrid Melbourne Mexico City Mumbai
Nairobi Paris São Paulo Shanghai Singapore Taipei Tokyo Toronto Warsaw

and associated companies in
Berlin Ibadan

Published by Oxford University Press, Inc.
198 Madison Avenue, New York, New York 10016

Library of Congress Cataloging-in-Publication Data
Asma, Stephen T.
Stuffed animals & pickled heads : the culture and evolution of natural history museums /
by Stephen T. Asma.
p. cm.
Includes bibliographical references (p.).
ISBN 0-19-513050-2 (alk. paper)
1. Natural history museums—History. 2. Natural history—Exhibitions—Designs—History.
I. Title.

QH70.A1. A75 2001
508'.074—dc 00-040674

Book design by Adam B. Bohannon

1 3 5 7 9 8 6 4 2
Printed in the United States of America
on acid-free paper

For my brothers, Dave and Dan

CONTENTS

ACKNOWLEDGMENTS

I am happy to acknowledge the generous support that many people provided throughout this project. Many museum curators, librarians, and staff people helped me in my research, and I am particularly indebted to Eric Gyllenhaal, John Bates, Doug Dewitt, Tara Hess, Nina Cummings, Shelly Ericksen, Simon Chaplin, Frédéric Guilbert, and John Wagner. I wish to express my gratitude to Christian Trokey, whose professionalism, skill, and good humor helped to secure the image permissions. Thanks also to Arthur Naiman, Peter Olson, Mary Kay Brennan, Audrey Niffenegger, Richard Ross, Allah from Milwaukee, Kathryn Koch, Jim Christopulos, Peking Turtle, Tom Greif, Baheej Kaleif, Les Van Marter, Sue Warga, Columbia College's Faculty Development Grant, the Lake Geneva Campus of George Williams College, and my students, who continue to teach me many important lessons.

My agent, Sheryl Fullerton, believed in this project from the start, and her careful attention to the manuscript and the editorial phase improved the book immeasurably. My excellent editor at Oxford University Press, Kirk Jensen, patiently guided the manuscript from its histrionic infancy to its current form. Of course, I take full responsibility for the book's remaining flaws. Stephen Jay Gould offered encouragement and advice at the outset. Thanks also to Michael Ruse, John Greene, Michael Ghiselin, and especially Phillip R. Sloan, who shed significant light on Cuvier's and Hunter's respective collections. I wish to express my appreciation for the ongoing support from Genie Gatens-Robinson and John S. Haller, and the very pleasant "Building Bridges: Religion and Science" conference at Southern Illinois University.

Other authors will understand me when I say that, in addition to the exhilaration involved, writing a book is a thoroughly manic and solitary experience. If not for my family, this project would have had me in the rubber room many times. So, my deepest gratitude to Ed, Carol, Dave, Dan, Elaine, Lynn, G-man, Maddy, all the Wagreich and Sherman clans, and, most important, Heidi, whose encouragement and support made much more than this book possible. When my anxieties were escalating unchecked, Heidi assured me that if I died in a freak accident, she would make sure that the manuscript was indeed published, but, alas, she would not honor my request to be preserved as a stuffed taxidermy mount.

INTRODUCTION

Toward the center of London's small Hunterian Museum stands an enormous human skeleton labeled the "Irish Giant." The giant skeleton was once the frame of Mr. Charles O'Brien, born in Ireland in 1761. O'Brien came to London in 1782 and exhibited himself, charging half a crown, as "the tallest man in the world" at 8 feet 2 inches. The towering man seems to have suffered from acromegaly, a pathological enlargement of the body that results from an overactive pituitary gland.

John Hunter, the famous British anatomist, met the Irish Giant in London and openly vowed to have the Giant's skeleton for his scientific collection. By 1783 O'Brien was on his deathbed, and, apparently thinking he should be less exhibitionist in death than in life and fearful that his skeleton would indeed end up in Hunter's collection, he paid some fishermen to ferry his soon-to-be-dead body out to sea and weight it down with lead. But Hunter, somehow informed of this development, intervened at the last minute and successfully bribed the ersatz undertakers. He brought O'Brien's corpse to his Earl's Court home, where he prepared the specimen himself by boiling it in a huge kettle.

Two centuries later I stood gawking in amazement at O'Brien's colossal skeleton. Compared to Hunter's nightmarish pathology cabinets to the left, the skeleton actually seemed quite tame. Contemplating these ghastly but fascinating exhibits, I felt as though I was learning more than scientific information. I was glimpsing the workings of John Hunter's mind, the mind of a man who seemed simultaneously a genius and a madman. It's no wonder that science has been saddled with Frankenstein stereotypes. Imagine living down the lane from John Hunter, where on any given day you might see him chasing his mutant roosters around the yard, or feeding his bright red pig, or ushering into his barn some shady grave diggers carrying mysterious bulky sacks.

The protagonists that you will encounter in this book are not only worthy of serious study because of their contributions to life science and museology, but also utterly intriguing characters. They not only collected curiosities; they were curiosities. Throughout this book, I try to give some palpable sense of the people who historically worked (and currently work) behind the scenes at natural history museums. I also spend a fair amount of time trying to unravel the cultural and theoretical convictions that museum exhibits create and reflect. Museums are saying more than we have previously noticed, and many of those messages stem from their history, their cultural context, and the assumptions that led to their formation.

Nature collections, in their institutional form, have been sources of education and entertainment for nearly three hundred years, and yet little appreciation exists in the public (and professional) mind for the philosophy behind exhibit displays. Museum presentations are three-dimensional windows into the world of ideas. But while observers successfully perceive and contemplate the factual tidbits that are placed in focus by the "window," they frequently fail to notice the presentation frame itself. This is not surprising, since the art of museum presentation, if it is to be effective rather than distracting, *should* be invisible. By reflecting on this previously invisible art of display and how it developed over time, I try to open up new facets of meaning for all museum excursions. If we learn the skill of reading between the lines at natural history museums, we begin to see deep ideological commitments quietly shaping and editing the sorts of things different cultures and different historical epochs consider to be knowledge.

My hope is that, after reading this book, museum-goers will not be able to look at exhibits the same way again. Ideally, this book itself should be a decoder device that readers can draw upon for future museum visits. Once you've been taken behind the scenes to witness the backstage drama that is present in all natural history museums, and once you've learned some of the historical and intellectual background of specimen collecting and exhibit creation, you'll find, I hope, every subsequent exhibit encounter to be more enlightening and more enjoyable.

The word *museum* was originally a Greek term meaning "place of the Muses." At the beginning of the *Iliad* and the *Odyssey*, Homer invokes the Muse of epic poetry to lend some of her inspiration to his literary portrayal. Obviously, I'm not comparing myself to Homer, but I will take any help I can get in order to tell my story. So in addition to Calliope (the chief Muse), I'd like to invoke Clio (Muse of history) and Thalia (Muse of comedy) to help tell this story. Invoking Muses here is very fitting, because natural history museums are not just places of information, but also places of inspiration. There is a psychology of museum experience and museum design, an affective character that has received little analysis to date. Hardly anyone would disagree with the notion that art museums are inspirational—who isn't uplifted and animated by great art? But the experience of the science-based museum can be every bit as inspirational, every bit as edifying as that of the art museum.

The main focus of the book is natural history. The collecting, investigating, and exhibiting of natural history specimens makes up the central subject matter of all the chapters. But it crosses into other areas, including anatomy, physiology, medicine, and even visual art, because the borders themselves have only recently been erected, and most of the book's historical protagonists did not recognize them as we do today. Throughout this century, biological science has been fracturing into increasingly specialized fields. Given the amazing advances and the Niagara of information, this fracturing and specialization have been necessary. For the most part, the days of the successful dilettante are over in the sciences. But the history of natural history—in fact, the entire development of life science, from the Greeks to the Victorians—is populated by quirky "all-rounders," who thought nothing of dissecting animals in the mornings, drawing plants and observing wildlife in the afternoons, and conducting surgery in the evenings. One of the reasons disciplines intertwined (and this is the issue most interesting to museology) is that the same specimen could serve several different purposes, potentially giving insights into medicine, natural history, anatomy, physiology, evolution, even God's goodness. Only toward the end of the nineteenth century did the sciences and the collections balkanize into the condition that we're more familiar with. So, in a way, this book is not so much a history of natural history as an intellectual memoir of the specimens themselves and the collecting cultures that managed them.

The odd thing about a specimen is that it's a kind of cipher when considered in isolation. Specimens are a lot like words: They don't mean anything unless they're in the context of a sentence or a system, and their meanings are extremely promiscuous. You can't gain admittance into the meaning of a specimen simply by looking at it harder, or even by anatomizing it. The significance of the collected object does not inhere in the specimen itself, but is socially and theoretically constructed. A bird's wing in a curiosity cabinet, for example, signifies something quite different from what the same wing signifies in a homology exhibit. A given specimen functions differently in a biodiversity exhibit than it does in a taxonomy exhibit. I've tried to track this contextualization or construction process in a variety of museums. Some thinkers from the humanities have been so taken by this theory of meaning that they've decided to make their own imaginations stand as the proper "signifying context" for meaning. This is not my approach. While I have occasionally included my subjective responses to collections, my main objective in analyzing these specific collectors has been to understand what the meaning was for them and their respective audiences.

Last, a word about style. Academic writers spend a lot of time trying to get themselves to disappear from their writing. In most academic writing, personality and voice are eschewed. So I guess that means this is not an academic book. I've previously engaged in this practice of self-effacement, but I elected not to do so here. If truth is the kind of thing that admits of degrees, then a narrative style is more true to the facts in this case. Moreover, the narrative has a vague gumshoe quality to it, and for this I make no apology. Anybody who's ever researched a subject (which is to say, anybody who's read more than two books on a given topic) knows that the exhilaration of the hunt is a material issue. It's all the more important to include the exploration itself, rather than just the findings, in a book that is substantially concerned with the process of museum learning.

In addition, I've tried to be honest about my own complicated interests in the museums and their strange specimens. My own excitement for this material swings like a pendulum between the gutter of morbid fascination and the ponderings of "pure" knowledge. I've included this for two reasons. First, it's a show of solidarity. My twisted curiosity can't be so much more pathological than that of John and Jane Q. Public. At least, I keep telling myself this over and over. Second, there is a strongly subjective character (emotionally, psychologically, and cognitively speaking) to any museum encounter. I came to learn, through my interviews, that museum curators and developers take these subjective aspects very seriously, so I thought I should take mine seriously as well.

This book draws on my research at various museums, including London's Natural History Museum (South Kensington), the Hunterian Museum (Lincoln's Inn Fields), the Muséum national d'histoire naturelle—Grande Galerie de l'évolution (Paris), the Galerie d'anatomie comparée (Paris), the Musée de l'homme (Paris), the American Museum of Natural History (New York), and substantial work at Chicago's Field Museum.

The Field and the American Museum of Natural History (AMNH) were obvious choices for exploration, since, historically, these have been two of the largest and most influential natural history institutions in the United States. Together they embody some defining aspects of American collecting and exhibiting. So too, London's Natural History Museum and Paris's Grande Galerie de l'évolution were chosen because they are the British and French flagship institutions and likewise embody their respective national orthodoxy. The smaller and more obscure collections of the Galerie d'anatomie comparée and the Hunterian Museum were chosen for two reasons. First, the curators of these European museums, Georges Cuvier (1769–1832) and John Hunter (1728–1793), were pivotal characters in the development of modern collecting, natural history, physiology, anatomy, and medicine. Their ideas about nature, and their methods of collecting and displaying, were highly influential throughout Europe and America. Understanding their respective museums allows us to understand practices of nature construction prior to the development of the theory of evolution. Second, these particular museums have been preserved, for the most part, in their original plan of organization. The rhetoric of the specimen presentation plan can still be experienced by today's visitor, some two centuries after the original exhibition. Needless to say, this historical preservation is incredibly rare and equally illuminating. The other collections examined in the book were chosen to illustrate specific theoretical points or cultural issues. The book is not intended to be an exhaustive survey of all modern museums.

Organized as a series of expeditions to and reflections on these U.S. and European museums, the book explores a network of interrelated issues. Chapters 1 and 2 are ostensibly concerned with the secret arts of specimen preservation. For example, why does the Field Museum in Chicago have flesh-eating beetles in its back rooms? Why did Peter the Great collect oddities—and what did he do with them? But while the preservation narrative forms a spine throughout the first two chapters, readers are also introduced to the basic thematic limbs of the book, including the origin of science museums, the different functions and motives of collecting and displaying nature, the institutional divisions of research and exhibition, and the phenomenological and psychological experience of the museum visitor. While the first two chapters trace the changing technical practices of museology, the third and fourth chapters move into the complex theoretical changes that marked pre-evolutionary collecting. Collecting and displaying natural objects is fundamentally an act of classification, and chapters 3 and 4 explore developments in modern European taxonomy. Chapters 5 and 6 analyze the practical and theoretical challenges of natural history museums in a post-Darwinian world. How are concepts such as biodiversity and cladistic systematics represented in today's museums, and how do contemporary museums juggle God and evolution? The final chapter draws together the various threads of museology by reflecting on the phenomenon of visual thinking. Here the reader is invited to explore the epistemology of museum learning and the uniquely affective character of edutainment.

STUFFED ANIMALS & PICKLED HEADS

CHAPTER 1

FLESH-EATING BEETLES AND THE SECRET ART OF TAXIDERMY

COLLECTING AND DISPLAYING natural history specimens is a more complex and dramatic activity than most museum visitors appreciate. The specimens themselves, for example, have intriguing and elaborate histories that largely go untold, because, unlike fine art objects, their individuality must be subjugated to the needs of scientific pedagogy. Generations of visitors at the American Museum of Natural History, for example, examined an Inuit skeleton as part of a general anthropology exhibit, unaware of the skeleton's own peculiar history.

In 1993 the American Museum of Natural History finally returned this particular skeleton, the remains of an Inuit man named Qisuk, to his descendants in western Greenland. Qisuk and other Inuits, including his six-year-old son, Minik, had been "acquired" as living specimens during an Arctic expedition in 1897. After polar explorer Robert Peary convinced the Inuits to return with him, the new emigrants found themselves housed in the basement of the American Museum of Natural History. Shortly after arriving, Qisuk died of tuberculosis, and unbeknownst to Minik, the museum staff removed Qisuk's flesh, cleaned his bones, and put him on display for New York audiences. Some time later Minik, who originally had been told that his father's bones had been returned home for proper burial, stumbled across his own father in an exhibit display case.

It was a similarly remarkable case that first drew me to explore the backstage stories of museology. I learned from one of my colleagues, a professor of Russian history, that Czar Peter the Great (1672–1725), whose motto was "I am one of those who are taught, and I seek those who will teach me," had a tremendous curiosity about the natural world and a penchant for collecting oddities. Peter's hoarding sensibility, together with a desire to share the accumulating discoveries of the modern age, resulted in one of the first public museums, in 1719. But the boundaries of what might be considered knowledge were drawn differently then, and freakish specimens were mixed chaotically with more sober displays. Reflecting on the oddities and curiosities in Peter's collection eventually raised interesting questions about how and why cultures deem certain things worthy of display. It's not entirely clear, for example, why some specimens are considered scientific and others merely lurid.

As the average Russian patron strolled through Peter's cabinet of curiosities, inspecting first the rare butterfly collection and then the exotic skeletons, she almost certainly would have stumbled across Foma. Foma was one of the "monsters" that Peter brought to the court as a living museum specimen. He wasn't really that odd, actually—just a human boy missing some digits. He had only two toes per foot and two fingers per hand (fig. 1.1). This somewhat common abnormality is still referred to as "lobster claw" deformity. It is a cleft between the second and fourth metacarpal bones, with the third metacarpal and phalangeal bones usually missing. The index finger and thumb fuse together, as do the fourth and fifth finger. These "blended" digits are oriented in opposition to each other and function like a lobster claw.

Apparently the novelty of Foma and other "specimens" such as a young hermaphrodite wore off quickly for Peter, who put them to

Figure 1.1. Foma, the living "curiosity" from Peter's museum. (THE ORIGINS OF MUSEUMS, EDS. O. IMPEY AND A. MACGREGOR, 1985, BY PERMISSION OF OXFORD UNIVERSITY PRESS)

work doing manual labor around the compound when the museum was closed. The hermaphrodite child eventually escaped from the curiosity cabinet. Foma, on the other hand, died young, and Peter had him stuffed for permanent display at the museum.

Despite a little embarrassment about the macabre interest, I have to admit that my first response to this story was: How do you stuff a human for display? Does one remove the innards, and if so, how is it accomplished? Is the stuffing made of cotton, hay, or what?

Peter took the collecting sickness even further (as if Foma weren't bad enough) by having some human heads preserved in jars of liquid. These weren't just any heads. He had William Mons, his wife Catherine's lover, and his own lover, Mary Hamilton, executed, and he ordered the jar with William's head to be placed in Catherine's chambers. Maybe he was thinking that he and Catherine could put their past infidelities behind them and start afresh.

Here again it was the preservation technique that puzzled me at first. Jealousy is no mystery at all, but pickling a head! Now *that's* a conundrum. In fact, fifty years after the unlucky noggins were first submerged, Catherine II, the wife of Peter the Great's grandson, was wandering through the back rooms of the palace and discovered the two jars high on a dusty shelf. She remarked to Princess Dashkov that the faces were amazingly well preserved and still displayed their youthful beauty. Catherine's sense of propriety must have overpowered her aesthetic appreciation, however, for she quickly ordered the jars to be buried.

Peter the Great had unknowingly bequeathed two quandaries. How are "dry" specimens, such as Foma, prepared for display, and how are "wet" specimens, such as the heads, prepared for display? Foma, one suspects, could not have been preserved by the embalming techniques of wet preparation, because he could not have been posed in a display position without an internal armature. The skeleton, one might object, *is* an internal armature. But of course the ability of the living body to strike a pose is a complex concert of muscle tissue, tendon, and bone, all forming the engineering miracle of internal leverage. It was becoming increasingly obvious to me that I needed more insight into this strange practice called taxidermy (fig. 1.2). What is the relationship between embalming and taxidermy? And why, come to think of it, have we felt the need, for many centuries now, to hoard dead things and display them in groups? Asking the "how" question about a stuffed Foma was leading me to the equally compelling "why" questions.

Things that previously seemed so obvious to me—so obvious that I never even contemplated them—were becoming increasingly

Figure 1.2. An amazing collection of taxidermy, some dating back to the eighteenth century, can be found crammed into a Paris shop called Deyrolle on the rue du Bac. This contemporary shop is a chaotic display that harks back to the curiosity cabinets of premodern Europe. (PHOTO BY AUDREY NIFFENEGGER)

strange. I've spent most of my life in the Midwest, and in that time I've seen countless deer, trout, sturgeon, owls, falcons, and squirrels all frozen in silent tensity on the walls of diners, homes, and businesses, not to mention the veritable ark of inert animals quietly haunting our museums. My perspective shifted; I felt as though I'd been in a Platonic cave all my life mindlessly watching a parade of stuffed animals, and now I finally had some questions.

Some things are so ubiquitous, so "natural," in one's experience that it takes a moment of self-alienation to point out their artificiality. This can be as trivial as the food phobias and preferences that our different cultures express (dog meat horrifies here but satisfies there) or as profound as the taxonomic categories by which different cultures organize their experience. Modern European culture classifies a bald eagle, for example, by categorizing its species in the genus *Haliaeetus*, in the family Accipitridae, but the Navaho categorize their animals by linking them to corresponding natural elements. So for the Navaho, the eagle is always conceptualized in tandem with the mountain, and in similar fashion the crane corresponds with the sky, the heron with water, and so forth. Or consider the pre-Renaissance habit of organizing animals and plants according to their common uses. Herbalists classified plants primarily along medicinal lines, with little consideration for morphological similarity. Consequently, the table of taxonomic relations reflected only the usefulness of organisms to human beings. Of course, this scheme is great for medicinal purposes, but taxonomy should also try to carve nature at its joints instead of just where convenient. Today this anthropocentric taxonomy seems as naive, to our scientific minds, as objectifying our prejudices by classifying spiders and snakes as "bad" and bunnies and dogs as "good." We'll examine these classification issues more carefully in the chapters ahead, because all collecting and displaying is also classifying.

There are rather different ways by which cultures and epochs situate their experiences in larger webs of significance or meaning. Which ordering systems are the "right" ones? Which cultural systems of organization correspond most accurately with reality? Why would the sight of Foma propped up in a case drive contemporary Americans shrieking from the building but stand as sober science in eighteenth-century Russia? The question of why people collect and display will eventually become paramount, but here at the outset, it is important to grasp how natural specimens are constructed.

Recently academics have become increasingly interested in the "culture" of science. Unfortunately, their understanding of culture frequently remains too rarefied, focused more on theory (strictly intellectual activity) than on the everyday practices of scientific culture. But to understand the culture of natural history museums properly, one must not be too dignified to enter into the morbid practices of specimen preservation.

I decided to ask a contemporary curator about Foma in particular, and about taxidermic representation in general. I arranged an appointment with Dr. John Bates from the Ornithology Department of Chicago's Field Museum. But in the weeks before our meeting, I took it upon myself to learn more about commercial and trophy taxidermy. Granted, lay taxidermy may not be educationally motivated, but the technical process of representation is quite similar, and animal trophies are interesting expressions of middle-class collecting impulses, displaying important continuities with early museology. To that end, I interviewed local taxidermists, studied professional and amateur textbooks, and researched the potentates of American commercial taxidermy, the Jonas brothers.

A wonderful text entitled *Complete Handbook of Taxidermy*, by Nadine Roberts, gives one a glimpse into the technology and culture of the trophy collector. The publisher of the handbook, Tab Books, advertises many other "popular books of interest," including *How to Make Your Own Knives* and *How to Test Your Dog's IQ*. The author of the taxidermy handbook makes it clear in the first few pages of the book that the chief virtue of doing taxidermy is its rugged individualism. "Taxidermy provides the opportunity for privacy. Perhaps the reason nobody will bother you is because they are afraid you will ask them to help skin a fish, but whatever the reason, you will be able to work alone."

Ms. Roberts, and others like her, are really just artists who are passionate about a medium that few of us understand. Witness her inspiring plea to the neophyte to maintain artistic authenticity. Apparently one can stuff animals with molded plastic (we learn more about this later from Dr. Bates), and this is Nadine's advice to one who is tempted to take the "easy route" of ready-made squirrel bodies available through mail order: "The goals of the taxidermist should be far beyond the application of these ready-made forms and supplies. He must continue to reach for the naturalness that is elusive, for the summit which will always be unattainable, never permitting himself to say, 'This is good enough. I cannot do better.' This is not to discourage the beginner in taxidermy . . . only to point to the possibilities that lie ahead for the taxidermist who wishes to do work for which he will be remembered and of which he can be proud."

My awe for the artistic integrity of the taxidermist only increases when I discover that the potential health hazards connected with stuffing animals include anthrax, Bang's disease, bubonic plague, cat scratch fever, rabies, Rocky Mountain spotted fever, ringworm, and sarcoptic mange.

I spoke with some of the members of the Illinois Taxidermist Association (ITA) just prior to their annual display competition in Effingham, Illinois. One of the previous winners of the small-mammal category told me that the general public fails to understand or appreciate the art of taxidermy, and the ITA is trying to increase the art form's visibility. I asked the members whether they saw themselves as artists, and all assented without hesitation. One member, a former winner in the single-reptile category, explained that rendering an animal's face to look "alive," in all its three-dimensional potential, is a painstakingly skillful process. Nadine Roberts explains, for example, that even removing the animal's face from the original skull is a meticulous craft.

> As you begin skinning the head, feel around the ears before you disconnect them from the head. Be sure to cut deeply enough so the ears will be intact on the skin. Work slowly around the eyes as well, feeling and looking at both the inside and outside of the skin as you work. Take care not to cut the eyelids or the tear duct area. These will show on the finished mount even when repaired if they are damaged much. Prepare to be patient as you reach the area around the mouth and nostrils. The skin is grown close to the bone and to the teeth in some places and you

will want to avoid cutting the skin, especially on the face where repairs are rather difficult to hide.

Long before it rose to the level of art form, however, the fundamentals of taxidermic practice were in place. When our ancestors were primitive hunter-gatherers, they developed the technique of tanning animal skins for clothing and shelter, and tanning remained a relatively low-technology, labor-intensive process for millennia. Things have changed recently because the naturally occurring chemicals (or engineered substitutes), including aluminum sulfate and sulfuric acid, are now easily available at drugstores and feed stores. The old-school practitioners generally practiced a method called bark tanning, although some hide preparation techniques apparently called for rubbing the brains of the animal into the skin.

First our predecessors stripped the bark off fallen trees and pounded it into a pulp. This was added to water until a soupy solution was achieved. Next, the skin was stripped off the animal and the "fleshing" process of scraping away all excess flesh, fat, and membrane was performed. If the skin was spread with salt and left to cure for a few days, the fleshing went much more easily. If they wanted the hair off, they created a bath of hardwood ashes (which naturally contain lye) and water in a hollowed-out log and soaked the hide until the hair pulled loose. Tanning proper is achieved by placing the skin in the bark soup for several weeks, occasionally stirring it. To finish the process, they submitted the skin to repeated stretchings and then lightly oiled it.

Our ancestors obviously used these skins as clothing and housing, but (if comparative anthropology can be trusted) they also draped the skins around rocks and mud piles to simulate the living animal. This mock animal was probably the focal point for hunting rituals, wherein tribes reenacted previous hunts or prepared for upcoming expeditions. These earliest versions of taxidermy, in which animals were re-presented to an audience, appear to be three-dimensional versions of the rather more famous cave paintings. Like the evolution of spoken language, these visual examples of symbolic communication may have been important in the history of human adaptation, because being able to re-create an earlier episode, or portray a future event, is a power that most other animals lack (or possess only in a very rudimentary degree).

Through further research, I began to grasp the state of taxidermy in the eighteenth century. At the time of Foma's re-creation, the methods of taxidermy were actually indistinguishable from those of furniture construction. In the eighteenth and nineteenth centuries hunters brought their prize kills to upholsterers, who sewed the animal skin together and stuffed it with cotton and rags. It was this rather crude form of taxidermy, resulting in some horribly misshapen trophies, that led to the phrase "stuffed animals." If this was, in fact, the manner in which Foma was displayed—stuffed like a pillow—he probably appeared much stranger in death than in life. Transform *any* human specimen, normal or abnormal, into a baseball mitt, and you'll have an exhibit that will turn the head of even the most jaded museum-goer.

Of course, the act of representing the living animal in museums and homes today serves different functions than it did for our hunter-gatherer ancestors and even our eighteenth-century predecessors. But exactly what those differences are seems unclear. Is the stuffed animal of modern culture, for example, a bourgeois attempt to partake in the majesty of the royal menageries? If one cannot afford living exotica, are stuffed exotica good consolation? And did preagricultural (or preindustrial) society, for example, need a set of symbols that conveyed the fragility of ecological relations? The recreated animals in our current museums often seem like sad cautionary tales of our exploitative tendencies. But this ecology message is a very recent function of animal representation, because our species has only recently become a serious threat to nature. Our stuffed animals mean something different than our ancestors' animal representations did. And yet one wonders whether there is some deeper continuity between the varied epochs and functions of animal collection and display.

Sigmund Freud was an avid collector of cultural artifacts, and the rare objects that surrounded him, from Italy, ancient Egypt, and Greece, framed him as a cosmopolitan man and a torchbearer of civilization. While he does not discuss the specifics of his own collecting mania, he does make some fascinating and humorous speculations about the psychology of collecting. For Freud, collecting is like many other adult activities in that it can be related back to one or more of the juvenile sensual pleasures. Freud sees his famous trilogy of sexual development stages—oral, anal, and genital erotism—manifested in the adult collector's passion. Recall that the pleasures of breast-feeding constitute the oral stage, the pleasures of retaining feces constitute the anal stage, and the pleasures of genital stimulation, obviously, need no further clarification. While it's a stretch to see the oral and genital stages realized in the collector's psyche, it's not so difficult to see the anal stage.

Freud claims that adult character traits, like the passion for collecting, are sublimations of these early and primordial forms of pleasure. Referring to someone as "anal" if they are neurotically neat and orderly is completely commonplace these days, so the connection here to collecting should not seem so bizarre. Freud says that adult "stinginess" is also a clear sublimation of the juvenile pleasure in holding in excrement. Whether one applies the epithet *anal* to someone in a literal developmental way or in a metaphorical way, it's all about dreck retention. If you're a stingy person, you refuse to relinquish and you strive to retain—two obvious qualities, Freud thinks, of the collector. The feces, according to Freud, are originally perceived by the child as an integral part of himself (a positive production), and only rigorous training convinces the child to part with them. Perhaps collecting is an unconscious attempt to re-create one's original unity, and the collecting of natural objects (such as stuffed animals) is an attempt to extend the boundaries of the ego and end the alienation of the postuterine experience. Unfortunately, the Freudian analysis, while undoubtedly clever, is too speculative and untestable to be of more value.

French philosopher Jean Baudrillard suggests a common psychological thread running through our instinct to collect, arrange, and display objects: "Surrounded by the objects he possesses, the collector is preeminently the sultan of a secret seraglio." Baudrillard claims that collecting and arranging is itself an exertion of power or dominance, one that is remarkably successful when compared to our attempts to dominate and control living things.

Baudrillard's comments about collecting and displaying objects in general seem a little more trenchant than Freud's, especially when we think about the taxidermic process of re-creating the unmanageable living creature (and the wilder the better) into the manageably inanimate trophy. He points out that "ordinary [human] relationships are such a continual source of anxiety: while the realm of the objects, on the other hand, being the realm of successive and homologous terms, offers security. Of course it achieves this at the price of a piece of sleight-of-hand involving abstraction and regression, but who cares? As Rheims puts it, 'for the collector, the object is a sort of docile dog which receives caresses and returns them in its own way; or rather, reflects them like a mirror constructed in such a way as to throw back images not of the real but of the desirable.'" So perhaps collecting is ultimately a power game, in which we seek to get mastery over someone or something, and the object becomes a reminder and demonstration of that mastery.

The Jonas Brothers company has been the dominant American taxidermy institution for the better part of the twentieth century, and in researching its work, I came to appreciate Baudrillard's bold interpretation of collecting. The brothers Jonas (Coloman, John, and Guy) started the famous Denver business in the 1910s, and the company (which still exists) did some early "stuffing" for the Field Museum (fig. 1.3). But the bulk of their clientele were wealthy hunters (fig. 1.4). Analyzing the company's own literature (their information and sales brochures) provides an illuminating glimpse into the culture of trophy taxidermy, particularly when one goes back to the 1965 catalog entitled "Game Trails."

The catalog design and photography have that familiar look of having been incredibly hip and cutting-edge in 1965 but seeming

Figure 1.3. Coloman Jonas, one of the founding brothers of the Jonas Brothers taxidermy business. Coloman founded the company in 1908, and many American natural history museums employed the Jonas taxidermy skills. (BY PERMISSION OF JONAS BROS. TAXIDERMY STUDIO, COLORADO)

Figure 1.4. A living room decorated with taxidermy, from the 1965 Jonas catalog "Game Trails." (BY PERMISSION OF JONAS BROS. TAXIDERMY STUDIO, COLORADO)

somewhat pathetic and outmoded in 2000. The catalog gives one the same feeling as visiting Graceland in Memphis and discovering that Elvis's high-tech mansion filled with top-of-the-line gadgetry has become the worthless flotsam of today's culture. When I opened the brochure to the index I found a large boldface preface:

> THIS IS A BOOK FOR MEN. Real Sportsmen . . . Real men who thrill to the outdoors. Real men who cherish memories of happy adventures. Real men who in their homes, re-live again, and share with others, the intense moments on the game trails, when they pitted their skills against the crafty and elusive animals of the wild. Enjoy with us the splendid results of successful hunts, shown on the pages of this book.

So I paged through and tried to enjoy "the splendid results of successful hunts." I discovered the beautiful deer-hoof lamp (fig. 1.5), the twin TV stools made from elephant feet, and the lovable "businessman's desk souvenirs." The souvenirs were cigar humidors and ashtrays made from rhino feet. And a "real man" also needs album covers and wastebaskets made of zebra hide (fig. 1.6). Of course, there were all the predictable bear rugs and deer heads, but there were also less popular mounts such as the walrus, the mountain goat, and the wolverine. Under each of these photos was a little informational blurb, and under the wolverine it read: "Wolverine. One of the

Figure 1.5 (LEFT). Lamp made from deer or antelope hooves. Caption reads: "Very attractive and useful ornaments or gifts can be made from the feet of game animals." (BY PERMISSION OF JONAS BROS. TAXIDERMY STUDIO, COLORADO)

Figure 1.6 (RIGHT). Wastebasket, picture frame, and miscellany made from zebra skin. (BY PERMISSION OF JONAS BROS. TAXIDERMY STUDIO, COLORADO)

few animals that seem to kill just for pleasure." The irony of this quip inside a hunting catalog seemed to have been missed completely by the brochure writers.

The Jonas Brothers company still operates in Colorado today, and on my request they mailed me their latest catalog. The new brochure is quite swanky—high-gloss, full-color, cutting-edge marketing. The quality of the taxidermy is outstanding. If you want to stuff a full-sized hippo these days, it will run you $23,500. Open-mouth poses require an additional $100. If you want sable antelope horns mounted on a wooden panel, it's only $215. And the "novelty items" are still available today at reasonable prices. But now, in addition to the ashtrays and bookends, you'll find scrotum pouches for $145. I'm not kidding. It's probably best to just move along quickly and not meditate on this new item.

Page after page of animal trophies, which made no pretense to educate, began to drive Baudrillard's point home to me. These were not scientific specimens; they were unapologetic reminders that the hunter is master. Of course, even if Baudrillard (or Freud, for that matter) is right about the deep psychological aspects of collecting and representing, it would be only the beginning of the interesting and complex story. He may be quite accurate when one reflects on hunters and commercial taxidermy, but can the same be said of the natural history museum's use of specimens? Our analysis must be careful not to paint everything with too broad a brush. Like discovering the "essence" of most any phenomenon, figuring out the essential quality common to all forms of collecting is only a preliminary step. When Aristotle, for example, figured out that the essence, or common quality, of all humans was being "featherless" and "bipedal," most people rightly slept through the discovery. As I eventually came to explore more and more collections, I began to see that the real stories were to be found in the *diversity* of representational acts, particularly when analyzed historically. I was increasingly drawn to discover the variety of ways that representational objects could be constructed and could communicate conceptual information and emotional experiences. There is no grand unifying theory of all collecting activity.

The stuffed Foma was himself a representational act, created by Peter for his Russian audience. What was the underlying agenda behind Peter's display, and what was the emotional and cognitive response of his audience? Before I could tackle these philosophical questions, however, I felt drawn to answer the practical questions. What were the display methods, the nuts and bolts?

The time had come to see an expert on these matters, someone whose bread and butter is taxidermy and museum display. On a brisk early April morning I went to meet with Dr. John Bates. Of course, I had been in the "front stage" area of the Field Museum many times, meandering through the various exhibits, but now I needed to access the "back stage."

I entered the main hall of the Field Museum and found it relatively unpopulated except for a brunch that was taking place in the roped-off center of the enormous nave. The fifty tables of well-frocked dignitaries were being addressed by a balding middle-aged man behind a podium. The main hall is so huge that every sound seems to reverberate, and the speaker was shouting into a microphone to be heard above the wall of noise created by clinking glasses, screaming kids, and blasting water fountains.

The fellow, who was shouting something about archaeopteryx fossils, turned out to be John W. McCarter, the relatively new president of the Field Museum. McCarter came to the Field from the business world. He formerly worked on capital investment, acquisitions, and marketing for the consulting firm of Booz-Allen and Hamilton. His appearance at the helm of the Field is a predictable extension of the trend to make not-for-profit institutions more like businesses, and this general trend makes many of the scientists and curators nervous about the future of the museum.

"Edutainment" is the idea that McCarter and other museum leaders around the country are invoking to pump new life (read dollars) into the natural history institutions. Over the last five years the annual number of visitors to the Field has remained around 1.2 million, and this static figure worries administrators. By making the educational function of the museum more entertaining, leaders such as McCarter hope to boost admissions. Many of the people that I met during my increasingly frequent visits to the museum—curators, designers, and developers—were very concerned that the cost of some of this increase in entertainment might be a decrease in scientific content.

But here, in the first few minutes of his speech, McCarter seemed to be talking—shouting, actually—about science rather than business. He was addressing an elite group of members who had distinguished themselves by donating to the Field for thirty years or more. This luncheon was partly a thank-you gesture and partly a financial petition. "We are going to be very fortunate in the fall," I heard him shout, "to have one of the six archaeopteryx fossils actually travel to the Field Museum from Germany."

The archaeopteryx is a crow-sized animal from the Jurassic period (around 150 million years ago) whose skeletal impressions were discovered in Bavarian limestone deposits. This animal is extremely significant because it is such a clear verification of evolutionary theory. Those contesting Darwin's discovery, from the nineteenth century right down to current creationists, have complained that if evolution is in fact occurring, we should find transitional forms (sometimes referred to as "missing links"). The creationists rightly point out that if birds, for example, are the descendants of dinosaur ancestors—as paleontologists contend—then there should be some evidence of an animal that is half reptile and half bird. The fact that we don't see these transitional animals—a pastiche of different traits—walking around has led many a creationist to a smug sense of victory. Unfortunately for the creationists, however, the archaeopteryx has crashed the victory party and delivered the dreaded evidence.

Though it is doubtful that the archaeopteryx could actually fly like a modern bird, it clearly possessed winglike forelimbs, and its body was covered with feathers (fig. 1.7). It also had a large furcula (wishbone), which is indicative of birds, and a bird's pelvic bone. Combined with these birdlike traits, the animal had the toothed jaws and the clawed fingers of a reptile. It also possessed the elongated bony tail and abdominal rib structure of reptiles. While the archaeopteryx may not itself be the precise ancestor to modern birds, it certainly represents a transitional or intermediate form between birds and reptiles. If Darwin is right, then we should expect to see these linking animals occasionally cropping up in paleontology, and, lo and behold, it turns out that we do.

Plaster casts of archaeopteryx fossils can be found in almost any self-respecting natural history museum, but to have the actual fossil come to Chicago was indeed impressive. In fact, when it enters the Field, it will be accompanied by an armed guard. Guards will stand next to the fossil twenty-four hours a day.

Figure 1.7. Archaeopteryx reconstruction. (COURTESY OF FIELD MUSEUM, NEG. NO. GEO81958)

McCarter, however, only mentioned the event in passing and quickly moved on to focus on another recent accomplishment, the "Dinosaur Families" exhibit. This edutainment exhibit of robotic shopping-mall dinosaurs was drawing big crowds of feverish parent-dragging children, and therefore it apparently warranted more attention than the archaeopteryx event. I guess we wouldn't want any science information or theory to slow down the flow of entertainment, the flow of money. I was beginning to feel snobby and superior about the degeneration of culture and all that when I remembered that it was a stuffed Russian boy who had originally aroused my supposedly refined intellectual curiosities. But I was not yet interested in the intricacies of museum politics and issues surrounding edutainment; all that would come later. For the present I had taxidermy on my mind.

Dr. John Bates is a thin, athletic-looking man with short gray-black hair; he likes to dress in T-shirts and jeans. John is a bird scientist and had agreed to show me around behind the scenes in the ornithology wing of the museum. The first thing that struck me when we entered the large double doors was that this part of the museum had a distinctly academic—perhaps even cloistered—feel to it.

"Many people," John explained, "don't actually realize that we're a fully functioning research institution here. I think there are something like seventy-five Ph.D.'s working on various research projects all around the Field. Quite a bit of good science is going on behind the scenes, but most people exclusively associate the museum with the public exhibits. In fact, only about one-tenth of the museum's collection is out on display at any one time."

"Well," I joked, "you guys are all secretly locked away up here, inaccessible to us common folk. You're probably hiding the truth up here somewhere, closed up where the rest of us can't get at it."

"Actually," he said with a laugh, "there's a great deal more public access here at the Field Museum than we had at my old institution, the American Museum of Natural History, in New York. For example, we just held our annual Members' Night celebration and had over ten thousand members roaming through these laboratories. So we're actually pretty good about public access. After all, it's in our best interest to educate the public about our research; otherwise they might feel hesitant about supporting it."

John explained the wider significance of his own particular research project, which involved genetic mapping of South American birds. Each Amazonian bird specimen that arrives at the Field has a small bit of flesh removed from it, and this genetic material is placed in tiny vials and preserved in a freezer. This material is then analyzed by electrophoresis, which is a fairly standard way of taking genetic "fingerprints." A DNA sample from a bird is mixed with substances called restriction enzymes, which cut the DNA strands at specific points, producing fragments of varying length. The sample is then placed in a special gel and an electrical current is applied. A positive electrode slowly draws the negatively charged DNA through the gel. The shorter fragments will travel farther from the point of origin than will the larger fragments. An image is made of the pattern, producing an observable fingerprint that can be compared with other samples to determine kinship among individuals or populations. John also employs the relatively new technology called DNA hybridization, which uses heat to separate and rejoin strands of DNA, thereby measuring the genetic distance between different species.

A significant consequence of John's work is the improvement of our taxonomic understanding. Taxonomy is the naming, describing, and classifying of organisms, but in practice it turns out to be much more difficult than it sounds. Classical taxonomy usually placed animals and plants in groups by the criteria of morphology (shape), so if two specimens had similar-looking body shapes or sexual organs, then they went into the same theoretical box. But things are much more complicated, because it is not obvious which organs or structures should be analyzed for similarity. If you examine daffodils and tulips, for example, you'll find that they have similar root structures and the same number of seed leaves, so you might group them together (as did the seventeenth-century naturalist John Ray, calling them "tuberous herbs"). But if you examine the structure, called the corolla, that encircles the stamens and carpels of these plants, you find that daffodils have a united structure, whereas the corolla of the tulip is composed of six petals. And this difference may lead you to group daffodils and tulips in different categories (as did another seventeenth-century naturalist, Augustus Rivinus).

Perhaps one could avoid this problem of classification by turning from structures to environments or ecological settings. In other words, if morphology is misleading, then one might group animals together based on their lifestyle. Rabbits and hares, for example, appear quite similar, but an examination of their lifestyles—rabbits live in burrows, and hares prefer the open fields—correctly indicates a species difference. Unfortunately, this method has problems of its own, and the early taxonomists who followed this lead often grouped whales and dolphins with fish, because they "act" like fish by swimming in the water. But, of course, whales are actually much closer to humans and other mammals than they are to fish. And while we're at it, bats aren't birds, either.

Suffice it to say that these are only a couple of the many theoretical and practical problems connected with the study of taxonomy, and most contemporary practitioners are pluralistic about their methodological approach. They often group things by combining morphological considerations, ecological comparisons, historical relationships, interbreeding criteria, and John's specialty, genetic analysis.

John uses the genetic fingerprints to help resolve some taxonomic disputes. For example, if one species of bird gets broken into two populations by a geological impasse of some kind, and the coloration of the two groups evolves differently over thousands of years, then they may appear to be two different species. Genetic analysis will reveal the underlying connection of the two morphologically distinct populations. As an analogy, imagine running into two strange creatures for the first time: a Chihuahua and a Great Dane. Physical characteristics would never lead us to place them in the same species, but genetic fingerprints would.

For thousands of years naturalists have been confused about how to classify the flamingo. Its long neck and legs suggest that it should be grouped with the stork family (Ciconiiformes), but its webbed feet, honking voice, and downy young have led many naturalists to group it with geese (Anseriformes). Finally, in the 1980s, genetic analysis proved that the flamingo is in fact much closer to storks than geese.

The importance of good classification is not simply obsessive neatness—everything in its proper box. It is not merely another expression of Freudian sublimation. The true significance of contemporary classification is that it gives us insight into the evolutionary relationships between organisms, or at least it's the starting place for asking evolutionary questions. Taxonomy after Darwin has become a window on the history of life.

But if one isn't terribly impressed by such abstract theoretical matters and, like a good American pragmatist, wants to know whether any of this backstage museum stuff can "bake bread," then it must be pointed out that John's work has value here as well. The genetic comparisons that John is doing in the hidden halls of the Field Museum may actually help save the flora and fauna of the Amazonian rain forest.

As most people are aware, the South American forests are being leveled, and it looks as if not even the pop stars can stop it. The principal players in that destruction are deciding whether it should be the eastern or western Amazon that gets the ax. If the east is settled on, then the west will be partly preserved in response to world pressure, and vice versa. John, however, has genetic evidence that reveals amazing diversity throughout the entire Amazon. The tendency of business interests is to see the eastern and western regions as biologically homogeneous, with the corollary that trashing one-half would diminish overall numbers but not significantly eliminate kinds of organisms. But John's research into genetic fingerprints has revealed that great variation exists both between and within Amazonian species, and of course it is precisely this wealth of variation (in any species, ourselves included) that provides for the adaptations of the future. Variation not only is valuable in this lifetime but also is the reserve bank upon which future species survival depends. So John and others like him are arguing that if South American forest harvesting is inevitable or unstoppable, then it should be done in small pockets spread evenly throughout the region. In this way we can preserve greater percentages of the dispersed genetic variation; we can maintain the genetic reserves.

"So your work up here," I asked, impressed with such practical value, "may have a major impact on world events?" This sort of thing may occur every day in the sciences, but we don't usually have such palpable payoffs in the humanities, so I was taken with the idea of research actually changing something in the external world.

John shrugged. "Well, whether the policy makers actually *listen* to my argument is very unclear, I'm afraid."

Through some of my own research, I later found out that Dr. Bates is a player in a debate that goes back to the mid-1970s. Biologists at that time were divided about the correct strategy for saving the Amazonian rain forest. Some argued that only very large reserves could stop major extinctions, while others believed that the smaller "forest island" scheme would maintain current diversity and protect against local catastrophes.

Thomas Lovejoy, a Smithsonian Institution adviser and founder of the Biological Dynamics of Forest Fragments Project, began to study the ecological impacts of the two strategies in Brazil. His team has been tracking reserves ranging from 2.5 to 25,000 acres, and the results have been interesting, albeit unclear. Many flocks of birds, for example, disintegrate in reserves of 25 acres or less but flourish in microforests of 250 acres. If the forest island technique creates islands that are too small, then edge effects start to cause problems. Edge effects are the different problems (e.g., death of flora and fauna) that result from a general loss of moisture. But interestingly enough, when the cattle ranches between forest islands fail, some

Figure 1.8. Mounted owl, in the storage rooms of the Field's Ornithology Department.

animals and plants start to adapt to the pasture areas in between their more traditional environments. This complex research program continues to work on the issue of viable forest-swatch size, and it continues to make important conservation recommendations. The current culture working behind the scenes at the natural history museum can be politically engaged and philanthropically motivated.

John and I walked down the dark entrance hall of the ornithology section, and while he surveyed the cabinets, I noticed a huge stuffed owl peering over our shoulders (fig. 1.8). The earthy colors were beautiful, and the convincingly realistic eyes reflected light in an eerie manner.

"This stuffed owl," I remarked, gesturing, "is one of the things I want to learn more about."

"Well, the first thing you should learn is that calling it a 'stuffed animal' is somewhat offensive. The proper way to refer to this owl is to call it a 'mount.' I'm not trying to be difficult or anything; it's just that this taxidermy method doesn't involve 'stuffing' of any kind." He led me to the first large metal cabinet and keyed it open, pointing inside while he pulled out a wide tray shelf. "Now *these*," he said by way of introduction, "are your 'stuffed' birds." Lying neatly inside the tray were rows of gorgeous birds of paradise, brightly colored specimens from New Guinea. Each species's plumage was more brilliant than its neighbor's, and the tail feathers of the males were elaborate. In addition to three different species being represented, several sizes of bird were laid out sequentially according to age. And each bird had a tiny handwritten, color-coded tag attached to its foot. We stood there alone in the dark hall, and this first drawer that he opened—only one drawer among thousands—seemed like a marvelous treasure to me.

He held up a large bird of paradise specimen that had a little tag marked "Wallace" dangling from its foot. The bird's back and wings were a rich brownish orange, the long feathery plumage that trailed behind was silver with hints of green, and the bird's head was a striking golden hue (fig. 1.9). This specimen had been captured and prepared by none other than Alfred Russell Wallace (1823–1913). Wallace, in a classic episode of synchronicity, formulated the principle of natural selection independently of his contemporary Charles Darwin. Laid up with tropical fever during the monsoon season in Borneo, the globe-trotting Wallace formulated his ideas into an essay that ultimately prodded Darwin into publishing *On the Origin of Species*.

If one wanted to be a Victorian naturalist of any distinction, one had to be well versed in the taxidermic arts. Long before Darwin wrote the *Origin of Species*, for example, he learned the art of taxidermy from a freed black slave who worked at the Edinburgh Museum in Scotland. Originally from Guiana, in South America, the taxidermy teacher, John Edmonstone, first set Darwin to dreaming about the exotic lands where his H.M.S. *Beagle* voyage would eventually land him. Darwin says of John that he "gained his livelihood by stuffing birds, which he did excellently: he gave me lessons for payment, and I often used to sit with him, for he was a very pleasant and intelligent man." This training paid off well, because some ten years later Darwin disembarked from the returning *Beagle* with over 450 stuffed birds, the now-famous Galapagos finches among them.

In the early days of the European collecting mania, the birds of paradise were originally believed by some to be "footless" animals,

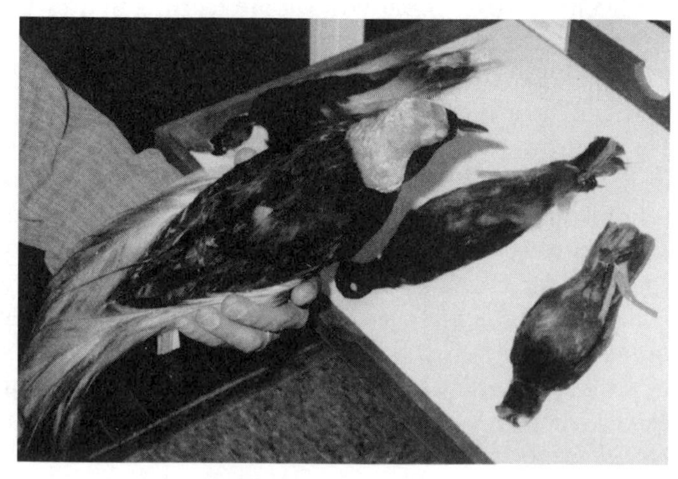

Figure 1.9. A "stuffed" bird of paradise that Alfred Wallace caught and tagged. John Bates holds the specimen in the back rooms of the Field Museum.

which simply buoyed themselves up by their flapping wings. A century before Wallace brought back living birds, the specimens were arriving in Europe with their legs already removed, so some classificationists drew the strange anatomical conclusion. Believing that the birds of paradise were footless creatures may seem remarkably gullible, but weird things were coming into Europe every day from previously unexplored territories. As it turns out, the native inhabitants of the Aru Islands, recognizing the trade value of the beautiful and rare bird skins, had begun supplying European traders with already prepared specimens. Since it was the brilliant plumage that the Europeans expressed interest in, the Islanders simply discarded the "superfluous" legs. Wallace was one of the earliest scientists to get deep inside the forests of the Malay archipelago and reemerge with many fully footed specimens.

Wallace spent over eight years in the East Indies, and for many of those years, after he and Darwin agreed to go public together, he amassed evidence of species transmutation. The preevolutionary understanding of animal distribution maintained that similar habitats were filled, by God, with similar animals. Since, the old argument went, organisms are perfectly adapted to their environments, searching through similar geographical environments should uncover similar organic bodies, similar physiologies, and similar behaviors. Of course, in reality, one does not find the simple correlation that preevolutionists predicted. The actual distribution made no sense prior to Darwin's and Wallace's discoveries. What one *does* find—tremendous diversity even in comparable habitats (e.g., the Galapagos)—is completely understandable in the light of natural selection. Successful groups of animals can migrate from their original habitat, and over time they become adapted to new surroundings. Sometimes there are impassable barriers to such migration, however, such as oceans and mountain ranges. Wallace argued that the radically different animals found in neighboring Australia and Southeast Asia were the result of a very deep waterway between Bali and Lombok, which effectively isolated the separate evolutionary pathways. Was this bird of paradise specimen, now in John's hand, a piece of crucial evidence for Wallace's eventual argument? Or was it one of his "throwaways" from a sack of two hundred such unlucky birds?

"These birds are fairly simple skin preparations, a couple of hours at the most. I do many of these preparations myself." He invited me to hold one, and it was extremely lightweight. He turned to me, trying to convey the nuance of his involvement. "It's remarkably therapeutic when you're giving your full attention to the preparation process. It relaxes you, puts you in that creative mind-set, and it provides you with a tangible end product to your labor." I tried to look as though I understood the Zen of stuffing, and I asked for more details of the process.

"Well, I like to begin with a slice at the bird's neck." He balanced the drawer and drew his finger across his throat, visually demonstrating all of the steps. I puzzled a bit at the fact that he demonstrated these cutting techniques on himself rather than on the birds in front of us. "You see, some people prefer to puncture in at the chest and then slice down to the vent, or some make an incision in the back—it's really a matter of personal taste. Now, I move from the neck down to the chest, and then carefully strip out the wing tissue and bone—it's almost like removing your arm from a tight jacket. But you can only peel so far. So with the wings and the legs, I have to pull the main fleshy bulk out but also cut the legs and wings, leaving some bone in the skin for support and realism. This is basically the same technique used on the bird's head, too. I pull the skin up—like someone pulling a tight turtleneck shirt off their heads—and when I get past the eye, I cut the head off and let the skin and beak stay attached to the skull."

"So," I interrupted, "the brain just remains forever in the skull of these birds? Doesn't that organic stuff just rot or something?" John smiled a little and turned his head to the side a bit, as if he was assessing and appreciating my interest in such undisclosed trade matters.

"No, you're correct. I have to pluck out the eyeballs and, working with a little brain spoon, I scrape out everything I can from the back of the skull. In general, as I'm peeling out the bird, I'm always sprinkling and rubbing dry preservative on the skin in order to cure it and soak up grease." I discovered later that cornmeal is a favorite absorbent used by taxidermists, especially on greasy birds such as duck and pheasant. And to protect against bugs, one can use alum or crushed crystals of paradichloride of benzene.

John continued to explain how he simply replaces the meat, viscera, and skeleton with a spongy foam or cotton stuffing, taking care to stitch it all together with a view toward anatomic accuracy. Lifelike accuracy, however, is not entirely necessary when the function of these hidden birds is to reveal scientific insights into coloration patterns, size comparisons, or gender differences—usually through statistical analysis of large numbers of specimens. These birds, though they maintain beautiful external markings, are really just bags of skin that have been emptied of their wet contents and refilled with lightweight, dry innards. But even these simple bags of skin speak volumes to the scientists.

By collecting and arranging large numbers of same-species birds of differing age, John and other ornithologists can determine the patterns of coloration and changes in markings that occur over life spans. These changes may turn out to have correlations with sexual maturity, camouflage needs, or important physiological systems. Of course, determining whether a bird is a youngster or an adult is the first step in arranging these trays, and this process turns out to be more interesting than simply looking at comparative body size.

Most people know that when human babies are born, the top of the skull, which is primarily composed of three bony plates, has not yet become one solid bone. At the fontanel, the soft spot on top of an infant's head, one can actually see the pulse, because the spot does not close until about eighteen months of age. (The human braincase actually continues to expand until a child is about seven years old.) When the fetus is moving through the birth canal, this malleable head comes in very handy, because it can stretch, bend, and squeeze in ways for which both mother and child can be thankful.

Very few people, however, are aware that birds have this same development of skull structure. An ornithologist can determine the age of a bird by checking the degree of cranial closure at the top of the skull. The morphological size of some species of birds does not radically increase over the lifespan, but this skull clue allows scientists an accurate assessment of age. And one can imagine how important this developmental marker becomes when scientists are encountering previously unknown or less familiar bird specimens.

The curious thing about the jigsaw bird skull is not that it reveals age but that it exists at all. If a malleable mammalian skull serves the purpose of getting the infant down the mother's birth canal, then why should a bird, which is already sheltered in a hard, protective egg, have a similar skull? If human skull structure is explained by reference to the function or "problem" of live birthing, then how do we explain the bird's structure when it has no such function or problem? Why is its skull in pieces? This skull similarity between myself and the bird of paradise that John showed me is what biologists call a homology. There are a great many homologies in the zoological world, such as the fact that a human arm, a mouse's forelimb, a whale's flipper, and a bat's wing all contain pretty much the same pattern of bones. And these homologies are particularly puzzling when the functions they perform are so different from one species to another. After all, why should a human arm and a bat's wing share the same blueprint when the two limbs perform radically different functions?

This skull riddle and other homology puzzles are created by our tendency to see adaptation in every biological trait. We post-Darwinians have all been trained to interrogate every anatomical structure with the question "What is it for?" Assuming that there must be some straightforward survival advantage for each and every trait turns out to be too simplistic. This is because some anatomical structures (and even animal behaviors) are helpful or harmless (neutral) remnants of some distant past ancestor. My skull is homologous with the bird's not because they are similar adaptations to birthing but because birds and humans share a distant vertebrate ancestor whose skull happened to have bony pieces. The different species that have branched off from that early ancestor have maintained the anatomical feature, some finding new and valuable adaptive uses for it (as in the case of mammalian parturition), others merely carrying it along as part of the developmental baggage. And since egg-bearing vertebrates (most fish, reptiles, birds, and amphibians) preceded live-birth mammals in evolutionary history, we can safely say that a jigsaw skull did not originate as an adaptation to the birth canal.

A homology, like the similarity between my arm and a whale's flipper, is different from an analogy—for example, the similarity between a bird's wing and a butterfly's wing. Analogies are structural or behavioral similarities that do not have a common ancestor. Both the butterfly and the bird wings perform the same function (allow for flight), but they derive from different anatomical elements. The wings look similar and function alike not because they share a historical connection, but because natural selection responded to similar environmental challenges with a similar adaptation (flight).

A fascinating feature of the genetic fingerprinting technology used by scientists such as John is that it can clear up the question of whether an animal trait is homologous or analogous. Old World and New World vultures, for example, share many traits: soaring food-search behaviors, powerful hooked bills, and featherless faces and heads. These similarities led taxonomists to assume a strong historical connection among the different species, placing all the vultures together in the order Falconiformes. But recent DNA analysis of the vultures has proven that New World vultures are related more closely to storks, and the similar traits found in Old World vultures are actually independently evolved adaptations to the carrion-feeding lifestyle. Some Old and New World vulture traits are analogs, not homologs.

This distinction between explaining a trait by direct appeal to adaptation and by appeal to historical ancestry became increasingly important when I eventually followed the trail to London, New York, and Paris. Explaining the inner logic of taxonomic similarity and diversity became a holy grail of sorts for eighteenth- and nineteenth-century collectors. Indeed, this impetus for museum representation is still very much with us. But more on that later.

The messages that dead animals can indirectly communicate, especially to trained senses such as those of John Bates, are remarkable. When the practice of collecting is infused by theoretical questions and principles, it transcends the mere will to hoard. Dead animals *do* tell tales. There are decoding secrets for interpreting the mute testimony of these animal specimens. Only in death do most animals pause long enough for our analytical minds to torture some truths out of them.

Once John Bates and other bird scientists have successfully organized their incoming specimens into gender and age categories, they can start reading some of this coded information. A nice example of this bird analysis is the McCormick Place collection program.

John opened a drawer of white-throated sparrows that, unlike some of the other species he'd shown me, all appeared the same in coloration and marking. All these sparrows were collected around Chicago's McCormick Place in the spring. I've lived in Chicago for most of my life, but the thought of McCormick Place as a site for scientific investigation was new to me. I had always associated the huge black exposition building, sitting on the lakeshore south of Soldier Field, with the annual auto show my dad dragged me to as a kid.

"What you're looking at here," John explained, "is a species of sparrow where the males and the females have identical shapes and markings, and subsequently they pose problems for scientists who are observing their behavior. However, by going around to McCormick Place every morning for four weeks in the spring and gathering the dead birds that have smacked into the windows there, we have learned some interesting facts."

"How many birds are we talking about, John? I mean, there can't be more than a handful of birds dumb enough to plow into McCormick Place."

"Oh, you'd be surprised," he responded. "Of course, some of it hinges on the particular weather patterns—heavy storms deliver up a bigger harvest of bodies, for example. But, generally speaking, over a hundred different species of birds are collected at McCormick every season. We have species of bird coming from as far as Argentina and Chile, and as near as southern Illinois." He could plainly see my amazement, but he refocused my attention on the sparrow drawer before us.

"We carefully record the time and day of each collected specimen, and as we do the skin preps we record the sex of the bird by checking the gonads. Then we chart all this accumulated information in the form of a graph, and it reveals an interesting pattern. When you step back and look at the data, it shows two distinct parabolas, one for males and one for females. It turns out that males are migrating to this area, on average, two weeks before the females arrive. Almost all the dead birds collected in the first two weeks of the returning season are males. These data finally confirm what we have long suspected but never had the stats to prove—namely, that the males always migrate first in order to set up territories. The males work out the hierarchies and the turf details before the females show up, and then they try to attract incoming mates with their respective spoils. And, of course, on the white-throated sparrows, which all look alike, we would have serious difficulty detecting this behavioral pattern by field observation alone."

So the drawer after drawer of skin preps accumulating in the back rooms of the Field Museum are not just for show in the exhibition halls. The labyrinth of preserved animals is more than just a repository for our visual consumption; it is a working, constantly growing database.

"Now this"—John swung around toward the mounted owl, remembering our original topic—"this is real artwork. You see, this owl is properly considered a taxidermy mount because we're trying to achieve a lifelike pose by using armatures or molded forms, but the birds of paradise are just skin preps, not taxidermy proper." John explained some of the subtleties of mounting large birds, and these basic techniques are applied likewise to mammals, reptiles, and any larger animal generally.

The first thing that curators do when an animal comes into the shop, so to speak, is collect as much data as possible. Exact measurements are taken of the body; photos and drawings are made in order to capture coloration and markings. Sometimes a death mask, which is a plaster or rubber cast of the animal's face, will be made for future reference. Next the animal is skinned, and the skin is salted and tanned. But the skinless body is also rigorously scrutinized and analyzed to record muscle configuration and overall subcutaneous morphology. After these reference data have been collected, the animal is entirely boned out, which means that the curator carves up and scraps everything but the skeleton. Then the skeleton takes a little trip to see the beetles.

When I looked particularly confused by his reference to the beetles, John motioned for me to step through some adjoining double doors (swinging doors reminiscent of the backstage portals of restaurant kitchens). We were in a larger, brighter lab room now, which had a few people working in the far corner, but John led me to a small, heavy metal door that looked like a submarine hatch. You know you're really backstage at the museum when you have to negotiate doors with air locks.

"Let me introduce you," John said as he opened the hatch, "to our dermestid beetles." I stepped into the foulest, most pestiferous stench you can imagine. A Dumpster outside a fast-food restaurant on a sweltering afternoon might re-create the fetid odor, sweet and sour in a nauseatingly pungent combination. Inside the claustrophobia-inducing room, which was significantly warmer, were about eight or ten glass tanks of varying size. The first thing that caught my eye was a large pile of fleshy bones. These large reddish black ribs had been recently boned out of an ostrich and were about to go into the tanks (fig. 1.10).

Inside each tank were thousands of dermestid beetles, otherwise known as flesh-eating beetles, blissfully chewing the meaty chunks and strands off the bones. Each bug was no bigger than a watermelon seed, but en masse they could strip a skeleton clean in two short days. This group was only *one* species within the heavily populated family Dermestidae, and I was reminded of biologist J.B.S. Haldane's quip, based on the fact that there are four hundred thousand species of beetles on the planet but only eight thousand species of mammals, that God must have an inordinate fondness for beetles.

John explained that the beetles could do a much more thorough cleaning job than the curators, but if they ever got out of the fleshing room, the entire collection would be at risk. The flesh-eaters would multiply rapidly with the prospect of such an expansive food supply, and it would be hard to stop the consumption of thousands of taxidermy mounts. In fact, the beetles are so voracious that the curators

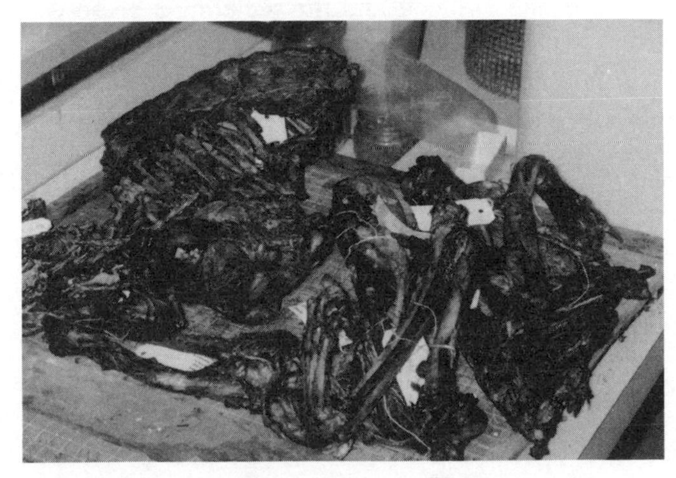

Figure 1.10. Ostrich bones, waiting to enter the flesh-eating tanks for cleaning.

have to time carefully the duration of each bone's immersion, or else the beetles will begin to dig into the calciferous material of the bone itself and thereby damage the skeleton.

Suffice it to say that witnessing this organic meltdown was repulsive but mesmerizing. Here was one of those quiet processes of degeneration and regeneration that go on interminably all around us but are rarely perceived. As we started out the door of this gory little dining room, I noticed a tiny black dot on top of one of the mesh-screened tank lids.

"Say, John," I blurted anxiously, "it looks like you've got an escaping beetle here. Should we do something? Quarantine, or something?" He glanced back, uninterested, and flipped off the light.

"Nah, don't worry about it."

After the beetles have picked the bones clean, the skeleton is rebuilt using wire and steel rods for support. This armature is positioned in the desired stance, and then an oil-based clay is used to sculpt the muscles over the top of the skeleton. Considerable artistry comes into play at this phase, but also significant anatomical understanding. Indeed, herein lies the major difference between the museum taxidermist and the commercial (or even hobby) taxidermist. Most commercial craftspeople, instead of custom-making each animal armature, order those premade polyurethane body molds that Nadine Roberts was lamenting in her taxidermy handbook. Ms. Roberts points out that deer heads can be ordered in a variety of lifelike positions: "For instance, deer head forms can be obtained for the following mounts: shoulder, neck, semi-sneak, sneak, and browsing."

Museum taxidermists hand-sculpt their clay versions of the animal and then make multipiece plaster molds of the model. The oil-based clay is then recycled for new projects and the skeleton is preserved for independent display or research storage. Then, from the plaster molds, a lightweight polyurethane or fiberglass manikin is cast. This manikin is fitted with the faceplate, created separately from the death mask, and the hard parts (horns, tusks, beaks, etc.). Next the manikin is greased with hide paste and the tanned skin is stretched over the form, like a glove over a hand. Sewing up the incisions must be done very carefully, with close stitches, and finally glass eyes and details in oil paint are added. After explaining the technique and showing me some relevant equipment, John invited me to the other end of the main lab to witness some people performing a simplified skinning process.

Here, seated at a low table, were two teenage boys—volunteers, as it turned out—laboring over a small rodent with bare hands and the occasional use of a scalpel. Both of them strenuously avoided eye contact with me. One boy looked a bit older, and I asked him to explain his methods. He turned his head up toward me, repositioned his glasses on his face, and resumed work without comment. John had left me on my own with these volunteers in order to take a phone call, so, undaunted by the cool reception, I pressed on.

"What kind or animals are you guys working on?" I ventured. The younger one responded to my series of questions with monosyllables, all the while refusing to look at me. They were working on prairie dogs. They were eighteen and nineteen. Yes, they had done this before. No, they had never skinned a larger animal . . . and so on. When I asked how much time they volunteer to the museum, the older one took interest and raised his head.

"I love coming here," he said, finally meeting my eyes. "I'd come here every day if I could. I really like the ornithology department, but I prefer the zoology hall because there's a little more variety there. Sometimes, since I have a membership, I just spend hours downstairs in the 'Life over Time' exhibit, memorizing the information. And they've got this cool Godzilla movie playing on some of the video monitors down there. It's actually *Godzilla Versus the Smog Monster,* but they don't show the Smog Monster, I just recognize the suit. You see, that film used a new Godzilla design—well, really, it was only the nostrils that were completely reshaped, and some of the fins on his back. Of course, Godzilla has no real resemblance to the *T. rex* or the raptors, but I've got this vintage plastic model at home from the nineteen seventies and it shows a stegosaurus together with . . ." It was as if I had flipped a switch on this kid. And while he rambled on, I noticed his friend starting to really lean into his work.

The elegant description that John had given earlier of the graceful skinning process now seemed to me a bit sugar-coated—or perhaps I was merely witnessing the awkward missteps of the novice. Whatever the case, the kid was pulling with all his strength to get the skin up over the head of this hapless fresh kill. He was breathing hard through his nose as he put his weight into the struggle, and while the affair up until this point had been relatively bloodless, the drops began to hit the underlying newspaper with some frequency. This was not the soothing therapeutic meditation that John had described, nor did it fit the cheery analogy to pulling someone out of a tight turtleneck sweater.

When I refocused my attention on the older boy, whose droning about Godzilla minutiae had suddenly stopped, I found him staring at me. He had a peculiar detached look about him now, as if he were sizing me up with a clinical eye. I imagined his thoughts as he analyzed me—*First I'd puncture in at the chest and then slice down to the vent, or maybe I'd begin with an incision in his back and then work toward stripping out his right arm. . . .*

I backed away slowly.

"All right, fellas," I mumbled, smiling, "I'll let you get back to work. Uh . . . John?" I called out. "John? Where are ya, buddy?"

On my way out, jarred by my suspicion that the kid wanted to practice taxidermy on me, I finally asked John about Foma. He laughed but could only speculate about the methods that the Russians employed. Probably, he suggested, a wooden armature, cotton stuffing, and a bark-tanned skin suit over the top.

In the end, John couldn't say for sure how Foma had been prepared, but felt confident that it was some combination of the mounting techniques he had shown me. Having come so far in understanding the preservation of dead things generally, my Foma fascination was now giving way to all things museological. I wanted to understand all the mysteries of natural history representation. I wanted to explore the hows and the whys of the process by which ideas become things (the curator's creative activity) and things become ideas again (the onlooker's interpretive activity). All museums have a front stage and a back stage; having been to the backstage area with John Bates, I was beginning to appreciate the manner in which knowledge itself is being consciously engineered from behind the curtain.

As I was leaving the Field Museum I stopped off in the Education Department (another backstage chamber) and perused their collection of suitcase-shaped dioramas. These are portable models of miniature ecosystems that the museum has been loaning out to Chicago-area grade schools for over eighty years. Each of the nine hundred dioramas is housed in a Plexiglas-fronted polished mahogany case, about 23 inches high, 25 inches wide, and 7 inches deep. Many of these well-worn exhibits date back to 1911, when Norman Wait Harris endowed the lending program, and it is fascinating to see the progression of display techniques from those early days to the present. The suitcase dioramas were originally delivered to schools by a truck specially designed to carry the unique cargo, and the dioramas were rotated every two weeks. Today, however, educators must go to the museum to pick them up and return them.

A 1928 brochure that outlines the N.W. Harris diorama lending program reveals that the idea of edutainment is not new. "Realizing that the impressions obtained in childhood are the most vivid and lasting, and that, to the child mind, knowledge is most welcome when its acquisition is sweetened with a flavor of entertainment—

something different from the routine of the classroom—those entrusted with the administration of Field Museum of Natural History have always aimed definitely to provide education, in its most attractive forms for school children."

The assistant curator on duty at the time was kind enough to show me the newest group of cases they were constructing. Most of them were dioramas of cultural anthropology; one on religious symbols, for example, contained various totems, fetishes, crucifixes, icons, statues, and so on. After inspecting several, I turned to him.

"What, no evolution cases for kids?" I asked with a smile. He and another assistant, working across the room, both erupted into laughter.

"Hell, no!" came the amused voice from across the room. The assistant regained his composure and turned to me.

"I'm afraid that's too controversial, politically speaking. We want to keep the Field Museum in a good light, you understand. Many of us have wanted to create some evolution boxes, but we have to be careful about how and when it's done."

"Yeah," continued the coworker, still huddled over his computer, "we might try to do some evolution suitcases after the millennium anxiety dies down. Dunno, really."

The cases educate and entertain by bringing a little bit of the Field to the people, so to speak. A marketing-minded cynic can also see that the program drags into the museum parents whose kids have nagged them into submission. This phenomenon is referred to by advertising directors as the "nag factor." Marketers everywhere try to maximize the nag factor when creating ads for kids because it translates directly into consumer activity. Can the same be done for museums, for knowledge? Is the nag factor a justifiable means if the end is wisdom or understanding?

The philosophy of advertising is very straightforward: Lead the consumer to your product via some form of ego massage. Convince the average person, for example, that his deep egoistic drive—his pleasure principle—actually includes Duff Beer as its ultimate goal. And since the ego always needs feeding and rarely pulls back from the table too stuffed to continue, advertising will always be with us. But should the rising museum movement of edutainment embrace this consumerist model? Should museums hook people by making exhibits more "consumable"?

Plato makes a useful distinction between the types of faculties with which human beings are equipped. The distinctions are really just helpful metaphors, but they're not any more suspect than the three-part structure Freud bequeathed to us. Plato divides up each person into reason, emotion, and appetite, and he argues that having our faculties arranged hierarchically in that order (through education) would result in a just state and a just individual. Justice aside, though, what's particularly interesting is the distinction between the appetites and the emotions; both have the power to cloud rationality, but they are not the same thing. (As an aside, Plato proves that reason and desire are separate things by offering the example of a person who, gripped with tremendous thirst, can nonetheless squelch it if her intellect believes the drink to be poison. Since it would be a logical contradiction to say that an entity can thirst and not thirst at the same time, he concludes that we have different psychic agencies at work within us.)

Though Plato's talk of different parts of the soul may be too antiquated for contemporary tastes, he nonetheless has pointed out a valuable distinction. There does seem to be a qualitative, not merely quantitative, difference between emotions such as wonder, awe, and moral indignation, on one hand, and the appetitive drives for ego satisfaction, on the other.

The appetites, seeking the pleasures of the flesh, are easy targets for the mercantile forces of our culture. The various vendors of the world know just how to call up our appetites and ask if they want anything, to which the response is always, "Yes, send something over." But there does seem to be this other emotional dimension to human psychology, not wholly intellectual and not wholly appetitive, that is both sensual and sophisticated.

Most nature museums—and this goes back to the curiosity cabinets of the Renaissance and early Enlightenment—do not really titillate the appetites, as in the case of consumer manipulation. The feeling of wonder, or the sensation of the marvelous, *is* emotional and can intoxicate, but, unlike the appetites, it has no obvious object or specifiable goal. It takes one out of oneself in a different way than mere appetite satisfaction. It draws one on, but one cannot really envision its terminus. Enthusiasm (the word means "to be filled with the gods") is an emotion that museums often engender, and it suggests that one momentarily loses oneself to something bigger—in a word, transcendence.

We tend to forget that natural history museums are also places of inspiration. Despite a prevalent stereotype of science as dispassionate stamp collecting, there is a romance of science, and museums are prime movers in generating that emotional experience. The late Carl Sagan, science popularizer extraordinaire, tried to convey the spiritual quality of his work: "In its encounter with Nature, science invariably elicits a sense of reverence and awe. The very act of understanding is a celebration of joining, merging, even if on a very modest scale, with the magnificence of the Cosmos." And this experience of transcendence is available to everyone—including atheists.

Harvard naturalist Stephen Jay Gould laments the museum edutainment movement that is mimicking the commercialism of the theme park world. He points out that "theme parks represent the realm of commerce, museums the educational world—and the first, by its power and immensity, must trump the second in any direct encounter. Commerce will swallow museums if educators try to copy the norms of business for immediate financial reward." Of course, he's right about this, but we must also recognize the role that emotions play in driving and informing our intellectual pursuits. And emotions are more effectively triggered by powerful imagery than by scientific prose. Museums need to avoid the crass commercialism of appealing to our appetites, but they must continue to address our emotional faculties. They must continue to be sensual places.

Museums would do well, however, to think about *which* emotions they should cultivate in their patrons. For example, frightening the hell out of people could be done easily, given the virtual-reality technology now in hand, but giving people an experiential encounter with the sublime (an awe-inspiring feeling of nature's diversity, for example) may actually lead individuals to do further research on their own. Scaring people is easy, but it may do nothing more than raise the customer's pulse, whereas giving them the sublime is more difficult but may help create the sense of wonder and edification that leads to the pursuit of understanding.

The edutainment issue in museums is not a simple one, and I don't want to take refuge in platitudes about the evilness of "mere" amusement. The reason that things are muddled here is that the same answer can be given to two different questions: How do you attract a consumer? and How do you attract a learner? In both cases, the answer can be spectacle. But there are different species that exist within the genus *Spectacle*, and it's worth trying to draw a few distinctions.

Some things are spectacles because they're odd. The human psyche functions effectively in the world of survival because it has impressive powers of pattern recognition. Think about the evolution of the mind. Our mind groups our perceptions into quickly recognizable patterns that allow for easier manipulation and faster reaction to our environment. Those ancestors better able to organize their experiences—better able to recognize patterns indicating danger, for example—were better able to leave like-minded offspring. Cognitive and perceptual familiarity with the environment frees humans to interact with that environment in ways that are increasingly complex and beneficial. Obviously, this is still true today. Imagine if you were amazed anew every time the sun came up, or if you were shocked and startled by every one of your sneezes, or if your own family's faces were strange again on every fresh encounter. We actually have glimpses into such minds, such as in the case of the disorder called prosopagnosia, wherein brain damage can cause a person to no longer recognize family members' faces, or in the case of being under the influence of LSD, when one can regress, babylike, to rapt enthrallment at the sight of one's own hand. If we had to attend to everything in the way we did when we first encountered it (or an entity like it), we probably wouldn't live to see the next totally amazing sunrise. The normal human psyche, with its ability to familiarize perceptions, has a relatively helpful knack for taking things for granted.

The spectacle of the odd momentarily breaks in on this orderly and familiar world. For example, you expected her nose to be there, but it's actually moved or missing altogether; you expected him to have five fingers, but he has only two; you didn't expect the female hyena to have a penis, but she does; and so on. In these cases, patterns are not recognized, or pattern expectations are suddenly railroaded into novel areas. Oddities force us to attend, and in a context that is not life-threatening, most humans find that experience to be pleasant. The lure of the bizarre is not always lascivious. Museums figured this out a long time ago.

In some sense, the educational value of any specimen (whether it's a dissected organ or a stuffed animal) lies in its power to extend illumination beyond its own individuality. A true specimen is a species representative rather than an idiosyncratic particular. This explains why we freeze the otherwise fluctuating and transient individuals of nature into static universals, but it also suggests that displays of oddities such as Foma do not fit into this framework. Foma is not on display as a specimen (an expression of the common features of humanity); he's on display as the ultimate individual.

Some things are spectacles because they're wonderful. There is a growing literature these days by scholars who examine the phenomenon of wonder and recognize its role in the various encounters between different cultures. There's something monumental in the experience that we call wonder—something slightly overwhelming, something Grand Canyon–esque. The object of wonder is certainly odd, but it is odd in an awesome sort of way. Arguably, even the very tiny things that we find wonderful (such as microchips) are still amazing in this overwhelming and monumental way.

Something can also be a spectacle because it recognizes or projects some fears, anxieties, or pleasures that already reside in the viewer. Using Plato's rather than Freud's version of the alienated self, a definite attraction to the grotesque exists for many people. We could use a Freudian way of explaining the phenomenon, calling it a "death drive" (Thanatos), and even point out that Thanatos and Eros are intermixed, but this is just another description of what we already know (that people are attracted to the morbid), not an explanation of it. Fears about death also make us very curious about it, and, metaphorically speaking, the smell of death is everywhere in a natural history museum. Come to think of it, the literal smell of death fills many of the backstage rooms as well.

A short survey of popular culture will convince anyone that we have a real lust for the spectacle of violence, the threat of death, the show of aggression, the glimpse of animal finitude. The macabre is sexy. But lest we assume a superior attitude and chalk all this up to the recent "illnesses" of a profligate pop culture, we should remember that artists (even some very rarefied exemplars of high culture) have been exploring the Dionysian underbelly for millennia. Museums have understood, and played upon, this drama for some time (though perhaps more recently than in previous ages, since they've become increasingly public rather than private institutions). When I eventually made a study of London's Natural History Museum, for example, the unambiguous highlight of the dinosaur exhibit (where young boys were shrieking with fetishistic glee) was a life-sized diorama of three *Deinonychus* dinosaurs ripping the throat out of a *Tenontosaurus*. Blood was everywhere.

These different kinds of spectacles both draw upon and create human curiosity. Educational and entertainment institutions meet in the common-ground territory of the spectacular. But some spectacles lead to something cognitive or reflective, and the hope of the educator is to facilitate that trajectory. There is a place in that trajectory for the odd, the wonderful, and the grotesque. But some spectacles, using the same spectacular launching pads of human curiosity, only lead back to themselves. The thrill-ride spectacle can be "managed" in such a way that it leads to more of the same, not contemplation and reflection. The spectacle itself becomes the commodity.

One cannot take a totally negative posture toward thrill tactics in museums because, when used intelligently, the melodrama of violence and bloodlust can give us all a safe field trip to the recesses of our reptilian brains. Many of the contemporary natural history museums pose stuffed predators (such as the lion) and smaller carnivorous dinosaurs (such as *Velociraptor*) in surprising locations: over doorways, leaping from around corners, dropping from archways, and so on (fig. 1.11). The animal's razor claws are set, the toothy

Figure 1.11. A dramatically posed *Velociraptor* seems to leap down on patrons in the Field Museum's "Life over Time" exhibit.

mouth agape, the brow furrowed. We enjoy this momentary reacquaintance with our prehistory, in which life was nasty, brutish, and short. We get an unthreatening encounter with the threatening. These mock-frightening displays highlight an interesting feature of the edutainment process, the psychological game of the diorama experience.

The museum spectacle, whether it be the life-sized taxidermy mount or the suitcase-sized diorama, gives us a glimpse of the psychology of representation and imagination. Appreciating the museum spectacle is somewhat like appreciating professional wrestling: You know the punches are fake, but you dupe yourself into emotional involvement anyway. Standing at the glass window of a life-sized exhibit of mounted deer, for example, brings on the strange oscillation that's required in museums (fig. 1.12). You must oscillate between knowing that it's a man-made construction and suspending your disbelief to enter into a play-along relationship with the display. Only by playing along a little and taking the representation for reality can you be transported to the quiet woods with foraging deer, or be "frightened" by the menacing dinosaurs, or be deep in the ocean with a giant blue whale, or watch birds fly overhead (fig. 1.13), and so forth. Your incredulous self has to give your credulous self a wide berth in order for the imaginative magic to occur. You move between the literal Gradgrind-mind, which acknowledges the stuffing and the artifice, and the imaginative Munchausen-mind, wherein the stuffing becomes viscera and the glass window barrier disappears (fig. 1.14).

Once the imagination has been introduced into the discussion, all the familiar arguments about mass media can be overlaid on the issue of museum edutainment. For example, the easiest way to effect the psychological oscillation described above is through the danger-

Figure 1.12. Life-sized display of mounted deer in natural habitat, Field Museum Chicago.

Figure 1.13. Mounted birds hung overhead in the dinosaur halls of the American Museum of Natural History, New York.

Figure 1.14. Mounted zebras at the Field Museum, Chicago.

fright-thrill tactic. Using wide-screen cinema, robotic dinosaurs, virtual reality, and such, you can establish a very convincing transition from representation to reality. But one could argue that it requires much more imagination to enter into the world of the suitcase diorama. This is the classic kids-had-more-imagination-when-they-listened-to-radio argument that people raise against television. It seems somewhat reasonable when applied to the museum, but it can get silly quite quickly. If the argument is valid, then wired skeletons require more imagination than stuffed animals, bones piled in boxes require more imagination than full skeleton poses, simply reading the descriptions of bones requires more imagination than looking at the bones themselves, and so on.

Dr. John Wagner, who is simultaneously a research associate in zoology and the biology specialist at the Field's Education Department, has no qualms about the edutainment aspect of museology. Wagner regularly dresses up as Charles Darwin and leads school groups, community gatherings, and Elderhostel parties through the Field Museum. His beard is real, he told me with a laugh, but the Victorian costume helps complete the theatrical picture.

"Museum education is naturally theatrical," Dr. Wagner explained. "As Darwin, I walk around with people and I tell them stories from the rich biographical material that we know. I explain certain exhibits in more detail, and I answer questions—anything from 'How did I analyze the Galapagos Islands?' to 'Why did I marry my cousin?' Years ago, when I first proposed the Darwin impersonation to my boss at the Field, he said, 'No way.' And I realized then that he was a man of small imagination. So I first started to travel to schools, supported by the Road Scholars for the Illinois Humanities Council, and eventually the success of the show made the Field loosen up and adopt the program."

Dr. Wagner was naturally drawn to this reenactment method because it was the way that he himself had grown to love and understand history. After he and a group of like-minded adventurers reenacted French explorer Sieur de La Salle's canoe expedition from Montreal to the Gulf of Mexico, he understood how dry facts could come alive through dramatic re-creation, and he vowed to do the same for natural history. He explained that many natural history museums are currently increasing their dramatic component. The Milwaukee Museum, for example, uses life-size manikins to show their chief entomologist climbing through the forest, engaged in field research. And the Minnesota Science Museum has a theater where actors playing Charles and Emma Darwin discuss God and evolution theory in a mock parlor setting. Clearly, then, many curators see edutainment as legitimate pedagogy.

The portable-suitcase dioramas at the Field Museum also dispel anxieties about the dangers of edutainment. These beautifully crafted representations of nature were indeed intoxicating, but in a good way. They were a feast for the eyes (they were sensual), but they were companions to the intellect, not combatants. These portable cases undoubtedly lured in admission receipts from duly nagged parents, but they also fired the imaginations of little kids who eventually became adult students of nature. These sensual representations of nature did not distract from intellectual understanding; they were a crucial part of the education process itself.

When museums are criticized for providing too much show and not enough substance, the criticism is frequently driven by an old cultural prejudice. And while there are very good reasons to worry about edutainment, a prevalent Western prejudice against visual imagery is not one of them. This bias is partly a legacy of the Enlightenment's hostility toward the senses.

Sensual experience was interpreted during the early Enlightenment period by rationalist philosophers such as Descartes as highly misleading and confusing when compared to the clarity and certainty of mathematical rationality. In the 1660s, for example, Robert Hooke—one of the first microscope designers—warned his scientific colleagues of the dangerous lure of "pictures" and cautioned against the use of images in science: "For the pictures of things which only serve for ornament or pleasure or the explication of such things as can be better described by words is rather noxious than useful, and serves to divert and disturb the mind, and sways it with a kind of partiality or respect."

It may be true that one can be more easily manipulated by imagery than by formulae, but many intellectuals turned this possibility into an absolute antagonism between reason and truth, on one hand, and appearance and falsity, on the other. Intellectual historian Stephen Toulmin points out that one of the crucial features of the modern worldview "was the hard-line contrast between reason and the emotions. This was not just a theoretical doctrine, with intellectual relevance alone: rather, from the late 17th to the mid-20th century, it shaped life in Europe on both the social and personal level." Toulmin goes on to claim that "calculation was enthroned as the distinctive virtue of the human reason; and the life of the emotions was repudiated, as distracting one from the demands of clear-headed deliberation." Couple this philosophical distrust of emotion (and imagery as its trigger) with Protestant worries about the dangerous intoxication of ritualistic shows (read Catholicism), and you've got the recipe for a long tradition of suspicion about art and spectacle generally. There was an antagonism in museology between gawking at disorderly collections and the intellectual consideration of "principles." So it was that many of the early nature museums of the eighteenth century were quite methodical in scientific principle but rather dull and unartistic to the eye. In this, the eighteenth-century museums were breaking with the previous ostentation of the seventeenth-century curiosity cabinets—which were wild, chaotic accumulations of everything and anything extraordinary. But more on that later.

Focusing carefully on the Field Museum dioramas, one can appreciate the emotional and potentially ethical power of imagery. On the front desk sat a small buffalo family arranged behind the Plexiglas of the portable case. The nuclear family was arranged with the male standing protectively over the female, who was resting peacefully next to her newborn calf. The painted plastic figures were positioned on a sloping mound of Western terrain, and the backdrop was a blue sky and a hilly range that had been painted to blend with the three-dimensional foreground. The male and female parents looked directly out at me with a calm dignity—or was it noble grace? Or perhaps only bovine stupidity? Whatever the case, this little diorama was clearly modeled on a slightly earlier and much larger exhibit from the Smithsonian's National Museum of Natural History.

In 1883 William T. Hornaday built one of the first lifelike exhibits for the National Museum, one of the first in all of America. The exhibit, called "Battle in the Treetops," displayed two male orangutans in a territorial fight, and it was an early attempt to break away from the dry taxonomic displays that previous curators had arranged for a scholarly audience. Hornaday followed up this popular success in 1888 with his "American Buffalo Group," and thereby began a quiet revolution in museum philosophy.

Curators in the late nineteenth century were becoming increasingly aware of ecological relationships, of the interconnections between plants and animals. The environment was not just a negligible backdrop for animal drama, but was inextricably mixed into the life of the animal and vice versa. But early-nineteenth-century museums were comparatively dry and monotonous collections of skeletal specimens, revealing nothing about the daily context or activity of the animals. Hornaday and other contemporaries began to construct "habitat groups," placing animals and plants together, posing animals in action with sophisticated taxidermic techniques, and painting exciting backdrops behind specimens. But this new aestheticism wasn't just art for art's sake, nor was it just spectacle for the entertainment of a popular audience.

The philosophy behind Hornaday's increasingly beautiful exhibits was essentially ethical. His highly realistic "American Buffalo Group," which he composed after a collecting trip to Montana for authentic materials, was designed to do more than just convey factual information to the scholar. His work set out to evoke an emotional response in the museum-goer, and it succeeded. Something of the sublime is conveyed in Hornaday's meticulous work, and it calls forth an emotion of respect. It tugs on some deeply felt but often buried appreciation of nature. These visceral emotions, somewhat lacking in the intellectual exhibitions of previous years, were the most immediate way to dedicate viewer's hearts to the philosophy of conservation. Images have an immediate moral power that discursive language seems to lack, and this fact was not lost on Hornaday.

In the early twentieth century Hornaday arranged beautiful photographs of his habitat groups, together with an anti-sport-hunting appeal, into his book *Our Vanishing Wildlife*. The moral power of good taxidermy is suggested by the fact that Hornaday actually influenced the passing of the Migratory Bird Treaty Act of 1918. When this wildlife protection treaty was being considered, Hornaday sent each member of Congress a copy of his powerful book. Like the little diorama cases from the Field Museum, Hornaday's book brought inspiring images of nature directly to the inexperienced urban world—in this case, politicians.

So exhibiting specimens is not the purely objective process that one might imagine. Most display designers understand that in addition to disseminating information, they are empowered to evoke important emotions through imagery. Curators are not without agendas, and displays are not without subtext. But who decides which agendas should be pursued, and why?

On the evening of September 15, 1943, a meeting was held in the theater of the Field Museum commemorating the fiftieth anniversary of the museum. The director opened the meeting by announcing that the trustees, in careful consultation with Marshall Field's grandson, were henceforth changing the museum's name from the Field Museum of Natural History to the Chicago Natural History Museum. (Apparently this name, nobly proposed to indicate public ownership of the collections, never caught on.)

After this introduction, the group of dignitaries presented speeches about past and future functions of museums. The fascinating part of these speeches, which I dredged out of the Education Department archive, is that the curators and directors are very self-conscious about their unique position in history. One of the luminaries present, Albert Eide Parr, then director of the American Museum of Natural History, in New York, sketches the development of natural history museums and then poses the question of where it was all heading.

Parr points out that the agenda of the earliest museums was simply to collect, to make a complete inventory of nature. Nature museums functioned as cornucopia-like displays of God's ingenuity and fecundity. But after Darwin, all these collections of natural organisms began to make sense from an evolutionary perspective: Some things were classed together because they were actually blood relations; some of these strange bones were evidence of extinctions; some of these bone sequences illustrated change over time; and so on. The second phase of museums, then, was to function as a warehouse of "proof" for evolutionary theory. The agenda of natural history museums in the late nineteenth century was to display, for the average admission-paying plebe, the persuasive hard evidence for evolution. And Dr. Parr claims that this pro-evolution campaign was successfully completed.

The Field Museum, originally the Columbian Museum of Chicago, came into existence during the third phase in the development of natural history museums. After evolutionary theory crossed from novelty status to established fact, natural history museums, with the good fight already won, seemed confused about their future function. In the early twentieth century nature museums, according to Dr. Parr, took a misstep. Instead of taking up the increasing challenge of educating the public about the need for environmental concern—about delicate ecological relationships—the museums tried simply to distract and entertain patrons with the exotic lions of faraway Africa or some other novelty.

When natural history museums turned to this function of "exotica merchant," they set themselves up for a demise that still may be imminent. By establishing an agenda in which patrons would flock to the collections to experience realistic taxidermic re-creations of exotic animals and faraway lands, the museums *seemed* to be choosing wisely. After all, where else could the average Chicagoan walk through mysterious jungles and deserts and gaze upon bizarre creatures? But the museums of the early twentieth century had underestimated their competition—indeed, had not even seen it coming.

We are now so absorbed in it that we forget that the twentieth century was the age of the photograph and the motion picture. How could taxidermy possibly compete with film images (and eventually television images) of exotic animals and faraway places? Dr. Parr points out that in 1943 the layman was becoming "fairly familiar with what he might expect to find beyond the horizons, from a multitude of other sources than museums." This chosen function of the museum was being usurped by other media.

Dr. Parr pleads with the Field Museum to change course with him and focus the dioramas and the displays on closer-to-hand ecology issues. Parr warns that museums cannot continue to play the role of exotica merchant, because people can buy better versions of that elsewhere. And likewise they cannot return to the preevolutionary function of collecting for the sake of collecting. He points out that "if we merely go on accumulating an ever more intricate inventory of nature, beyond anybody's needs or capacities to use, then the gratitude we can expect will be as slight as the service we render."

Over fifty years later the Field is still trying to define its function, and these debates over future missions are eerily resonant with Dr. Parr's earlier musings. Recently, for example, the Field Museum delivered up a remarkably lame exhibit entitled "Dinosaur Families," which consisted of shopping-mall dinosaur robots that squeaked and jerked about in a manner so gripping that my niece fell asleep as we watched it. The designer of this particular exhibit, who did a noble job with poor materials, confided in me that patrons actually spent much more time engrossed in the tiny display cases of archaeological tools than with the lurching robotic dinosaurs. I can see the ghost of Dr. Parr shaking his head at the Field Museum's continuing attempts to sell us the exotic. How can such feeble exhibits as "Dinosaur Families" compete with Spielberg's *Jurassic Park*? It's not that the museum should stop trying to fire our imagination; to the contrary, I believe that this is one of its greatest powers. But it cannot do this by the same methods as mass media.

The good news is that the Field Museum has been increasingly, albeit unknowingly, following Parr's recommendation. In a 1997 essay in *Museum News*, the then director of the Field, Peter R. Crane (who recently moved to a directorship at Kew Gardens), claimed that the most pressing issue for the institution was to educate the public on environmental conservation issues. To that end, the Field has recently established an Office of Environmental and Conservation Programs, "to coordinate and focus the Field Museum's environmental collections and research programs across departmental lines, and to help strengthen the linkages among the academic departments and between our academic and educational programs."

Indeed, my own experience of learning about Dr. John Bates's ornithology work is powerful evidence that a vital relationship can exist between the different agendas or functions of the natural history museum. The academic research of collecting and analyzing specimens is a function that directly feeds another museum function, namely, sensitizing the public to biodiversity and conservation issues.

While the challenges for the Field and other such museums are daunting, I feel hopeful that they will adapt. Good taxidermy mounts may become a thing of the past (just as stuffed boys such as Foma have disappeared from museum collections), but new media (virtual reality, no doubt) may be successful in igniting that same peculiar curiosity—part sensual, part intellectual.

As I began to make my way toward the southern exit I stopped for a while at a compelling diorama of grizzly bears. I contemplated the craftsmanship that went into these mounts—the poses were realistic, and the stitching was invisible. This life-sized arrangement of animals reminded me of the tiny buffalo family that I had seen earlier in the suitcase dioramas. Here was a baby bear nestled up to a kind-eyed, nurturing mama bear, with a strong protective papa bear towering over them, ready to fight off any potential threat. I stood there for a few minutes, contemplating this natural nuclear family—and letting the museum, the place of the Muses, work its magic on me.

Eventually I learned that this "natural nuclear family" was indeed a bogus construction of nature by curators (see chapter 6), but here, on my first encounter, I was captivated by the animals' enormous size. Then I began noticing the unique qualities of their fur, the size and shape of the claws, the distribution of their body weight, and so on. One of the values of taxidermy, at least originally, was its power to slow down, to actually freeze, the creature long enough for our perceptual equipment to register the details. It is precisely this immobilized quality that makes all taxidermy inherently creepy; not only are the cubicle-bound creatures alienated from their real environment, they're also alienated from the distinguishing property of life itself, which is motion. A lot of things in nature either move more slowly (e.g., plants) or more quickly (birds, etc.) than we can properly track, so their movements are, practically speaking, invisible to the human eye and mind. Before photography, the mounted specimens brought morphological details and subtleties into focus for scientists and laymen alike. Photography, obviously, was able to freeze the lightning-fast aspects of nature, and "moving photos" were able to speed up those aspects of nature that were slower than the snail's pace. But even before these time-management technologies, there was taxidermy.

CHAPTER 2

PETER THE GREAT'S MYSTERIOUS JARS: HOW TO PICKLE A HUMAN HEAD AND OTHER GREAT ACHIEVEMENTS OF THE SCIENTIFIC REVOLUTION

A FEW WEEKS LATER I was on the trail again. I had learned a tremendous amount about the techniques, the history, and the philosophy behind taxidermy, but I had not yet gained an understanding of "wet" specimen preparation. I now understood the basics of a dry preparation such as Foma, but my mind returned to Peter the Great's disturbing jars. Recall the czar's strange display of the decapitated lovers. It turns out that Peter's pickling of his paramour's pate was a very early application of some new preserving technology.

This time my curiosity had brought me from Chicago to London. My wife, Heidi, and I took the tube from our hotel in South Kensington to the Holborn stop, and we made our way down Kingsway on a brilliant June morning. We eventually found ourselves in the lush greenery of Lincoln's Inn Fields.

The park, entered through a large mock Tudor archway, was quite calm compared to the bustle of Kingsway's business district. Wigged barristers hurried to and from their offices, lawyers arrived to play tennis during their lunch break, and homeless men slept off hangovers on nearby benches. Strangely enough, this bucolic park was employed as a public execution site in the seventeenth century.

The beautiful Lincoln's Inn, which served as the Court of Chancery in the mid-nineteenth century but dates back to the Shakespearean era, stood in the southeast corner of the park. The Georgian architect and collector Sir John Soane had his home and museum at the north end of the fields, and the neoclassical Royal College of Surgeons (RCS), built in 1836, loomed to the south (fig.

Figure 2.1 The facade of the Royal College of Surgeons, Lincoln's Inn Fields, London. The college houses the Hunterian Museum.

2.1). It was the Royal College, or more specifically a collection inside the College, that I had come from Chicago to see.

Housed inside the RCS is one of the greatest collections of wet preparations in all of Europe, and indeed the world. In the late eighteenth and early nineteenth centuries, European museums became heavily populated with this new kind of specimen. After uncovering the hidden art of taxidermy, I had set about researching the wet prep method and in that process realized the need for a pilgrimage to this relatively obscure collection. Prior to arriving in London, my investigations had revealed some fascinating information on general postmortem preservation.

The story of wet preparation is tied up with the development of both embalming and anatomy. A few comments on these topics are necessary, then, before we address specimen preservation directly. Embalming is the procedure with which most people are familiar, so it seems a suitable place to begin. But while the practice is indirectly familiar to us, it is also (like other taboos surrounding death) situated in terra incognita.

Most Americans are clueless regarding the details of the embalming procedure. There is a remarkable paucity of public information on postmortem preservation. But many Americans have stared down at some dead relative resting in their open casket—and for that matter, most Americans will themselves end up the same way, while their relatives stare dumbly down at them.

Jessica Mitford's classic book *The American Way of Death* (originally published in 1963 but revised by her in 1998), acknowledges the public's blind spot concerning preservation techniques: "Embalming is indeed a most extraordinary procedure, and one must wonder at the docility of Americans who each year pay hundreds of millions of dollars for its perpetuation, blissfully ignorant of what it is all about, what is done, how it is done. Not one in ten thousand has any idea of what actually takes place. Books on the subject are extremely hard to come by. They are not to be found in most libraries or bookshops." Indeed.

My father had actually worked for a while as a part-time mortician. He explained to me that the first procedure in cadaver preservation is the draining of all blood through the jugular vein. Next a product called Flextone is pumped in through the carotid artery. This fluid is mostly formaldehyde, but it also contains pigment (for skin coloration) and perfume. After you have a few gallons of this mixture circulating through the cadaver, my father explained, you usually sew the mouth closed and glue the eyelids shut. Then, using a tool called a trocar, part needle and part gun, you vacuum out the internal organs and pump cavity fluid into the gut. Before you can begin the cosmetic work, you have to wait for eight to ten hours while the tissues firm up.

My father was somewhat hazy about the restorative procedures that came next, but with the help of Jessica Mitford's book, together with Strub and Fredrick's definitive text *The Principles and Practices of Embalming*, I gathered the cosmetic details.

If the undertaker was not able to drain the blood shortly after death, a series of swellings and discolorations might ensue. These unpleasant developments pose creative challenges for the restorative artist. If the mouth has swollen up, Strub and Fredrick recommend cutting out tissue from inside the lips. If the artist accidentally removes too much, the facial shape can be restored with cotton padding. If the neck becomes swollen, internal tissue can be removed through vertical incisions carved on both sides of the neck, and you can rest easy that when the deceased is casketed, the pillow will hide the neck sutures. You can't be too careful, though, and as an extra precaution against leakage, the neck sutures may be painted with liquid sealer.

All of this raises interesting questions about the theatricality of our society's death practices. The funeral is itself a dramatic performance, or presentation ritual, designed to elicit a variety of responses in its audience (e.g., edification, consolation, catharsis, etc.). The deceased person is represented or staged in a particular way by the funeral director to figuratively take a final bow. In fact, sociologists have pointed out that funerals, like museums and restaurants, have both a back stage and a front stage. Clearly, the gruesome procedures that my father witnessed at the funeral home are examples of the backstage dimension of death presentation. And this backstage area is strictly off-limits for the funeral audience, in much the same way (and for some of the same reasons) that a restaurant's kitchen is concealed from the diner: "[T]hose who have worked in the backregions of a restaurant, in the kitchen or after hours, and have participated in the preparation of food, the classification of customers by the staff, and so forth, may have difficulty seeing the frontstage performances in the same manner again." This distinction between front stage and back stage is something we've already encountered in the museum setting, and we will find later (chapter 7) that the distinction itself is being slowly dismantled in some contemporary museums.

Even the language used in the backstage area is uniquely different from the reverential murmurs of the front stage. The front-stage "loved one" is frequently dubbed with other nicknames while in the backstage area. Those who are unlucky enough to drown will find themselves affectionately referred to as a "floater," while a man who burns to death will be christened "Mr. Crispy," and the whole embalming process is frequently referred to, behind the scenes, as "pickling" or "curing a ham." This irreverent fact seems rather disturbing, but it's also funny. Comedy, among its various virtues, does seem to have a psychosocial function to it. Laughter is often the great release valve that allows us to distance ourselves from stressful situations; sometimes this is healthy, and sometimes it's pernicious.

Few or none of the emotional and intellectual audience responses at a funeral drama can be achieved if the boundaries between backstage and front stage are not strenuously enforced. As Erving Goffman points out, "[I]f the bereaved are to be given the impression that the loved one is really in a deep and tranquil sleep, they will have to be kept away from the area where the corpse is drained, stuffed, and painted for its final performance." Behind that classic beatific corpse smile, for example, is laborious backstage violence. Notice here that there is some analogy to the backstage violence and front-stage "artwork" of the museum specimens.

Now, this unappetizing, albeit fascinating, information is all fine and good if you're planning a very brief and passive funeral display. But the undertaker's artwork, while impressive, holds together for only a few days at most. Decay sets in very quickly because a relatively light ratio of preserving fluids must be used if the flesh is to appear lifelike. Funerary embalming, then, is still too impermanent for creating museum specimens. But the other area wherein postmortem preservation received considerable attention was the development of anatomical dissection. Anatomy and natural history were not completely autonomous disciplines during the late eighteenth and early nineteenth centuries. The inner and outer structures of an organism could shed light on an animal's behavior, and vice versa. In addition, natural history collections and medical collections were usually housed under the same roof. The museums that developed during the Enlightenment period could not effectively realize their educational function without well-preserved anatomical specimens.

In my undergraduate days I gathered some limited experience with postmortem preservation. I had studied anatomy when I was an art student, drawing cadavers in various stages of dismemberment. But the entire course was more macabre than educational (fig. 2.2). A handful of art students were allowed into the premed cadaver room, supposedly to learn anatomy while drawing the various corpses, but even the teacher—a scientifically illiterate illustrator—couldn't pretend to know his humerus from his radius. It was all too overwhelming for us. After all, we were artists, not scientists. We were preoccupied with our own tortured souls, expressing ourselves with grave self-importance through hackneyed symbolism on overwrought canvases.

It was strange, and perhaps a little bit sad, to realize that countless men of science had risked life and liberty to secretly carve up cadavers, trying to unlock life's mystery, while slackers like me could just pay their tuition and gawk at the grisly spectacle. The study of anatomy, from the ancients through the moderns, has always struggled with cultural and religious prohibition.

By the time of the Enlightenment—a bright period for empiricism—most European countries had established some laws to protect the medical study of human cadavers. It took an additional century, however, for England to make dissection legal. Actually, in

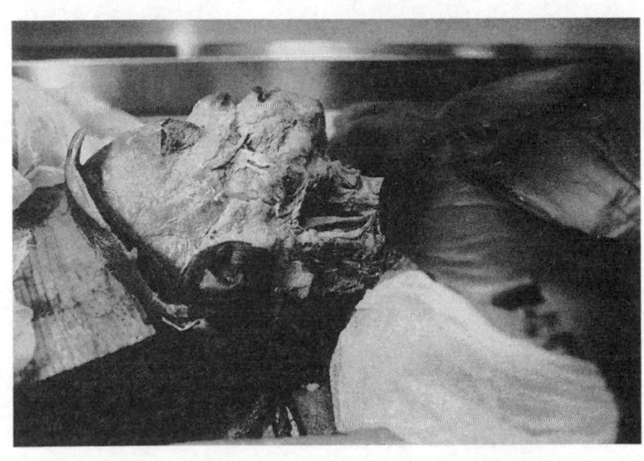

Figure 2.2. Profile of cadaver's head and neck, skin dissected away.

England it was permissible to dissect humans, but they had to be the bodies of executed criminals. There were considerably more medical students than executed murderers in the mid-nineteenth century, so a black-market grave-robbing trade emerged. The grave robbers of England became known as "resurrection men."

It was the case of the Scottish entrepreneurs Burke and Hare that finally persuaded Parliament to pass the Anatomy Act of 1832. It turns out that William Burke and William Hare developed a lucrative system of luring old men to their home, strangling the life out of them, and selling their bodies directly to the unquestioning medical school.

After Great Britain passed its Anatomy Act, Pennsylvania passed its own form of the law. This act stated that if a corpse had to be buried at the state's expense (because of the poverty of the deceased), then it first could be dissected by medical students. An anatomy professor at Philadelphia's Jefferson Hospital named William Forbes was instrumental in getting the act passed in 1867, arguing that unless the city decriminalized the process, Philadelphia would end up with a cadaver black market and a boom in "resurrection" vocations. But the act applied only to Philadelphia and Allegheny Counties, and soon the cadaver needs for Philadelphia's huge medical student population outstripped the legitimate supplies. In the early 1880s significant numbers of cadavers were being pilfered from the African American Lebanon Cemetery, and when a team of black citizens and journalists ambushed the resurrection men one night, they found that the trail of stiffs led back to William Forbes at the Jefferson Hospital. These events resulted in the passing of a statewide law in 1883.

Some people object in general to the brutal acts of cadaver dissection. Oftentimes the objection is religiously based, but sometimes it stems from visceral (pun intended) aesthetic revulsion. In these days of computer modeling, we have a digitized Adam and Eve floating serenely in cyberspace for all medical students to study. The dead bodies of a man and woman were frozen solid and then shaved into razor-thin slices by the U.S. National Library of Medicine. These slices have been CT-scanned and MRI-recorded, and then reconstituted in digital space for Internet immortality. With that kind of teaching aid available, one might ask why medical students continue the brutalization of corpses.

But brutalization just may be the crucial feature of the traditional anatomy lesson. If I'm in dire need of some serious surgical procedure, I'd rather not have a squeamish surgeon who's too sensitive for the nasty trench work. William Hunter, the eighteenth-century

British anatomist, pointed out to his students that "Anatomy is the Basis of Surgery, it informs the Head, guides the hand, and familiarizes the heart to a kind of necessary Inhumanity."

One of the authors that I came across in my research seemed almost *too* comfortable with the "necessary Inhumanity" of dissection. The author of an odd text entitled *The Story of Dissection* takes a singularly strident tone. In each chapter he works himself into a rant about the tremendous stupidity of all people who oppose dissection, eventually lamenting the "profound dormancy of the creative spirit in all human spheres, a spirit hopelessly mired in the viscosity of inert ignorance."

Throughout the book there is a strange defensiveness about the practice of vivisection. Vivisection is the act of dissecting "living" humans and animals; in the case of humans this usually meant convicted criminals. From a physiological perspective, experimenting on the living body, with all its fluids and pumps in action, is much more educational than dissecting inert bodies. It would be a glaring understatement to say that this is a painful process, unless one believes that wine is anesthetically sufficient for having one's limbs and viscera removed. The author of *The Story of Dissection* is annoyed that vivisectionists of bygone ages have been saddled with moral odium. He repeatedly claims that vivisection is not the barbarity that it seems to be, and we should not be derogatory toward these brave researchers "who had the courage of their convictions" to perform the very informative physiological analyses. He argues that vivisection is "the very serious means to a very noble end" (i.e., medical advancement).

After reading this treatise in the course of my research, I set the book down on the table, unsettled by the complex moral issues of such medical research. The author's tone was disturbing, but then again, the healing arts really did profit from such violence. My eyes happened to rest on the spine of the book, and I noticed that the author is listed as Jack Kevorkian, M.D. Unbelievable. Of course, Dr. Jack Kevorkian—aka "Dr. Death," as he's affectionately known in the United States—is the celebrity advocate for physician-assisted suicide and euthanasia. He has assisted in the suicides of dozens of terminally ill people, and he has been charged several times with murder, ultimately being convicted of second-degree murder on March 26, 1999, in Pontiac, Michigan. I discovered that Kevorkian was around thirty-one years old when he published *The Story of Dissection* and that while he was doing a residency at the University of Michigan Medical Center, he drove to a penitentiary in Columbus, Ohio, and tried to convince some death-row convicts to undergo vivisection experimentation. One of the convicts actually agreed to the proposal, but Jack's attempts to rally support among medical colleagues eventually cost him his job at the university.

One of the most interesting points that emerges when one surveys the history of dissecting is that the vast majority of early dissecting was done in cases of suspicious death or as a means of explaining a specific death. In other words, they were autopsies. Unlike Kevorkian's medical justification for dissection, the majority of early dissections were motivated by the pursuit of justice. In the early days one did not explore the anatomy of the dead to shed light on the health of the living. Given that resurrection men were still working 150 years ago, one can appreciate how late medical dissection made its way onto the scene. So the general trajectory of human dissection was from autopsy to medical anatomy and physiology. Animal dissection, however, has been practiced for so long, relatively speaking, that a kind of "pure" science arose there more quickly. Anatomical analysis of animals was not originally perceived as useful to human life or health; it was not obviously a practical science at first. Aristotle said that "all men by nature desire to know," and he, along with many other like-minded inquirers, opened up a lot of dead animals simply to understand their inner machines. Dissecting served, for this group, as a tool for "internal" natural history.

One can form a rough taxonomy of dissection motivations. There was the autopsy motive, the medical health motive, the pure understanding motive. And, of course, the dissections undertaken by artist-scientists such as Leonardo da Vinci constituted a fourth motive, one where anatomical understanding served as handmaiden to accurate aesthetic representation.

When I think back about my own brush with artistic anatomy in college, I remember the strong smell of formaldehyde. The pungent smell of it infiltrated my clothes during class and stayed with me well into the evening. My old college roommates became well acquainted with that sweet disinfectant stench every Tuesday and Thursday.

On those days, if I arrived early, I would help raise the dead. Each body, four of them altogether, was kept in a giant metallic coffin. Inside the large steel tomb the body lay submerged in a vapor of formaldehyde and phenol, and I would turn a large steering-wheel gear to slowly open the casket and raise the occupied platform. Each body was at least a year old, and each sprawled in a different stage of dissection. Bob (of course we named them) had a splayed leg off. Sue's chest had been turned into a leathery dermal vest, with easy open and close for heart and lung explorations. Most of the dissections were done by medical students on Mondays and Wednesdays, so our uncomplaining specimens were always dismantled from week to week in new ways. Al suffered a remarkable transformation late in the term when he rose from the formaldehyde soup with his skullcap sawed cleanly off, a sponge stuffed into his braincase. The following week Al's head had been buzz-sawed into two perfectly symmetrical profiles that nicely exposed his sinus cavities and bloated tongue.

This all sounds quite disturbing, but after the first day the cadavers seemed completely nonhuman. These weren't fresh dead. All their body parts had become a uniform taupe color, and they lay before us in an inert way. These were not taxidermic works of art. Al and Sue and Bob would not last long, and their brief altruistic afterlife could have been maintained by someone as untutored as myself. The preservative technique employed on my cadavers (an intensified solution of embalming fluid) would never sustain them as good collectible specimens.

It was precisely because this pickling method was so distorting and temporary, and only for this reason, that the American polymath Charles Willson Peale (1741–1827) never exhibited embalmed people in his museum. Peale established the first true museum in America, at the end of the eighteenth century (fig. 2.3). It was in Philadelphia that he amassed a huge collection of natural and cultural curiosities, ultimately expressing a desire to permanently exhibit great men's corpses for the inspiration of patrons. Friend to Benjamin Franklin, George Washington, and other luminaries, Peale felt that preserving their bodies for display would pass down high moral qualities to future generations. People could, Peale suggested, gaze upon these material remnants as representations of the lofty characters that once inhabited them; the intention was to inspire ordinary people to strive for equally lofty goals. The only problem for Peale was that the preservative techniques of the time were not entirely up to the task. A human face could become quite disfigured by the harsh preservatives, and subsequently the subtle effects of inspiration would never be achieved. It's a little bizarre that this display objective seemed appropriate to Peale in the first place. But a British contemporary of Peale's, a dedicated collector, had perfected the art of wet preparation to such a high degree that his pickled human faces remain subtle and "animate" to this very day. His work is what brought me to the Royal College of Surgeons.

In 1793 the extremely industrious curioso John Hunter died, leaving an unprecedented collection of 13,687 painstakingly prepared biological specimens. The collection was purchased by the British government and entrusted to the Royal College of Surgeons as a national treasure in 1799. The Hunterian collection is one of the

Figure 2.3. Charles Willson Peale, *The Artist in his Museum*, 1822. (PENNSYLVANIA ACADEMY OF THE FINE ARTS, PHILADELPHIA)

most scientifically important and aesthetically unusual museums in the world.

John Hunter was born near Glasgow in 1728, and in 1748 he followed his brother William to London, where they set up a private anatomy school in Covent Garden. At the school, John did the grunt work of preparing specimens for his brother's lectures, but his skills very quickly advanced to the level of art. John was a genius at dissection and preparation.

In 1761, during the Seven Years' War, Hunter was posted to Belle Île, off the north coast of France, as an army surgeon. Between saving people's lives and amassing research on surgery for gunshot wounds, he developed a treatise on geological formations and paleontology. When Hunter returned to London he began his own anatomical school, practiced surgery on the side, and was elected a fellow of the Royal Society in 1767. His interests ranged over every facet of natural history, and he procured many exotic specimens, eventually arranging the collection along theoretical principles for teaching purposes. Hunter's research became widely respected, landing him appointment as surgeon extraordinary to the king (1776); membership in the Royal Society of Gothenberg (1781), the Royal Society of Medicine of Paris (1783), and the American Philosophical Society (1787); and appointment as the surgeon general to the army (1790).

Hunter correctly diagnosed himself as suffering from arterial disease, and he occasionally suffered attacks of angina. Adjacent to his consulting room he kept a couch where he could repose when he felt an attack approaching, and he frequently told his friends that his life was in the hands of any rascal who chose to irritate him. At a contentious board meeting of the St. George Hospital on October 16, 1793, one of Hunter's fellow board members *did* choose to irritate him concerning a dispute over student privileges. This argument brought on a massive attack, and Hunter died just outside the meeting room.

After Hunter's death, his amazing collection was entrusted to the Royal College on the condition that catalogs of the specimens be created, a yearly lecture series be inaugurated, and the museum be kept open for extensive hours. Generally speaking, Hunter was an important transitional character who helped raise surgery from its former status as barber-shop bleeding to systematic science, and who helped move collecting from mere curiosity to the status of empirical education.

When you enter the Royal College under its classical colonnade, you ascend a huge spiral staircase, lined with portraits of famous surgeons, to the Hunterian Museum. Although opened to the public only in 1995, scholars and medical researchers have had access to the collection since its 1963 reconstruction (it was bombed in 1941). But the real fights over public access occurred in the mid-nineteenth century.

The conditions attached to the sale of Hunter's collection—catalog and library creation, lecture series, open admission policy—went largely unobserved, and the collection was still not open in 1827. The Royal College's closed-door policy led to riotous opposition from the scientific and political radicals in London. The nonaristocratic scholars and students of the university system, together with the working classes, perceived elitism in the Royal College's control of information. Imagine the indignation, for example, if the Field Museum of Chicago closed its doors to the public and allowed only scholars from Ivy League schools to visit.

The radical anatomist James Wardrop's complaint of 1825 typified a rising political tension surrounding Hunter's collection. Wardrop claimed that the Royal College's fellows "take our money, give us ipso facto laws, lock up our property, insult us with mock orations, live at our expense," and yet, he added, the people are as "entitled to the museum, and the property of the college, as any member of the court." Wardrop continued with a roast on a Royal College director:

> Some years ago Sir William Blizard promised that the library should soon be opened. We are still outside the door however,

and a part of the bust of Sir William, like the molten calf of the Israelites, may go down our throats, before we shall see a book, especially the Hunterian manuscripts.

There was more than just new information at stake, however, in these early disputes over Hunter's collection. Hunter's specimens, and particularly the way in which he chose to display them, became a focal point for a debate about the underlying principles of nature.

The entire museum is only the size of a basketball court, with most exhibits contained on the first floor and a narrow balcony circling above. There are no mechanical dinosaurs here, no hands-on interaction, no leaping dioramas competing for attention. In fact, if you didn't slow down and study the cases carefully—often having to turn on each case's lighting—you might altogether miss an exceptionally powerful experience.

One of the first displays that I studied was a jar containing a bleached white sack with two organic hoses rising out of the top; the bottom of the sack was eaten away (fig. 2.4). The bleached organ seemed to float suspended in a crystal clear fluid, but on closer examination I discovered two almost invisible threads holding the specimen in place. The exhibit, I learned, demonstrated the postmortem effects of gastric juices on the walls of a man's stomach. Hunter had addressed the Royal Society on this new discovery:

> There are very few dead bodies in which the stomach is not, at its great end, in some degree digested. These appearances I have often seen; at first I supposed them to have been produced during life, and was therefore disposed to look upon them as the cause of death; but I never found that they had any connection with the symptoms. It struck me that it was from the process of digestion going on after death, that the stomach, being dead, was no longer capable of resisting the powers of that menstruum, which itself had formed for the digestion of its contents. These appearances of the stomach after death throw considerable light on the principles of digestion, and show that it neither depends on a mechanical power, nor contractions of the stomach, nor on heat, but something secreted on the coats of the stomach and thrown into the cavity, which there animalises the food, or assimilates it to the nature of the blood.

We take for granted that stomach acids break down our food, but here is Hunter actually working it all out. Here he is laboriously crafting a piece of truth to which we are so accustomed that we never

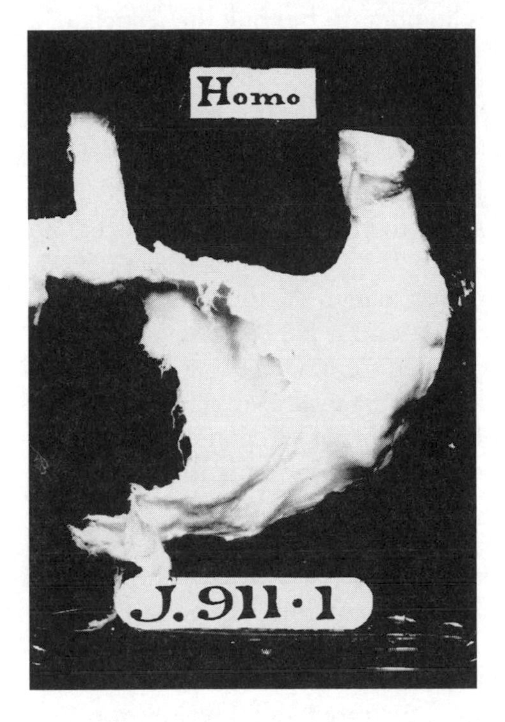

Figure 2.4. Wet specimen of human stomach, disintegrated by digestive acids. (BY PERMISSION OF THE ROYAL COLLEGE OF SURGEONS, ENGLAND)

pause to reflect on it. We have to imaginatively think our way into a world where food digestion *could be* the exclusive result of a mechanical pulverizing stomach before we can appreciate Hunter's genius. Such imaginative acts are necessary to appreciate anything from the history of science.

This bit of digestion wisdom, like all of Hunter's work, was based upon penetrating observational skills. Hunter represents the flip side of the early Enlightenment's hostility toward the senses. The rationalist distrust of experience was counterbalanced in the last half of the eighteenth century by a strong empiricism. But once the art of observation had passed through the crucible of seventeenth-century skepticism, it became much more principled and theoretical than the previous age of visually intoxicating "wonder cabinets." Hunter wanted visually arresting images not for their own sake but because they would give way under methodical study to the principles that underpinned all anatomical and physiological phenomena. If he did not find empirical evidence naturally, as in the case of the digested stomach, he went out and generated it. Hunter could not always wait for nature to yield up its secrets; sometimes he forced its hand a bit.

A great example of Hunter's experimental mind-set could be seen in a glass case containing four shelves of grafting experiments. Here were examples of homografts (same-species transplants) and heterografts (different species). For his homograft, he took the testicle of a cock and attached it in the belly of a hen. He patched up the hen (no more mention of the poor male) and forgot about it for a few months. After a while John killed the hen, dissected the gut, and found the testicle growing on the intestine. Now, if someone had told this to me, I wouldn't have believed it, but there it was before my eyes. Not only did it seem reasonable to Hunter to try it once, but he admits to trying it many times.

I later discovered that a biologist named Arnold Berthold conducted the same experiment in 1848, and it led to the eventual discovery of hormones. Hunter concluded from his own grafting experiments that the "living principle exists in several parts of the body, independent of the influence of the brain, or circulation . . . and in proportion as animals have less of brain and circulation, the living power has less dependence on them, and becomes a more active principle in itself." Berthold, on the other hand, was not interested in these "vitalistic" implications of a testicle transplant. Instead, he noticed first that castrated roosters plump up into fat and tasty adults (capons). Operating on the hunch that testicles must somehow interact with the nervous system or the blood to produce anatomical and behavioral characteristics, Berthold castrated two juvenile roosters and transplanted their testes into their gut cavities. Instead of becoming fat and inactive capons, the roosters continued their traditional aggressive male behavior. When Berthold dissected the birds, he discovered that an extensive network of capillaries had forged a communication between the testicles and the circulatory system. This eventually led to the discovery of those inhibiting and stimulating chemical messengers known as hormones. Hunter's strange testicle transplant seems bizarre at first, but you can never tell where science will uncover some important truth. The science of endocrinology was born in the belly of a rooster.

The next specimen shelf contained Hunter's heterograft. This time Hunter took a freshly pulled human tooth, bored a small hole in the crest of a cock's head, and worked the root end of the tooth into it (fig. 2.5). (What a cruel fate it must have been to be born a rooster on John Hunter's farm.) When Hunter finally killed the rooster months later, he sliced the head and tooth symmetrically down the center and discovered that the blood vessels of the cock's comb were now feeding the living tooth. Hunter writes of the experiment "that the external surface of the fang adhered everywhere to the comb of vessels similar to the union of a tooth with the gums and sockets." And, similar to other comments he made about many of his investigations, he confesses, "I may here just remark, that this experiment is not

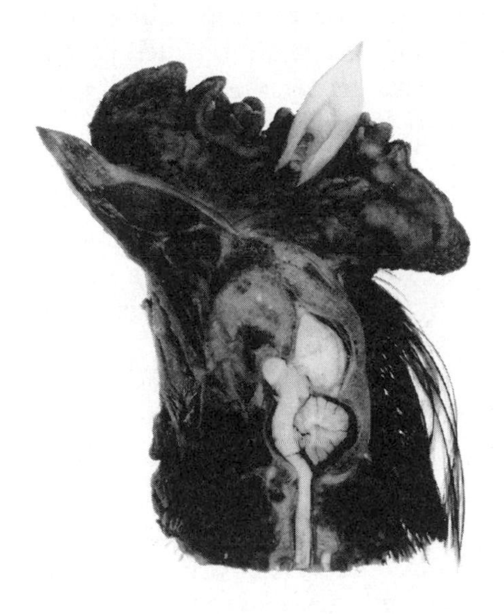

Figure 2.5. Cutaway profile of rooster head, with human incisor transplanted into the comb. (BY PERMISSION OF THE ROYAL COLLEGE OF SURGEONS, ENGLAND)

generally attended with success. I succeeded but once out of a great many trials."

Trying to better understand the relationship of structure to function, Hunter experimented on a gull's digestive system. In a cylindrical jar, the bird's splayed stomach floated before me. It had been floating there for two hundred years. The stomach walls, normally fairly thin, had become thick and muscular because Hunter denied the bird its natural diet of fish and instead fed the gull on grain for over a year. The change in diet led to a noticeable change in stomach structure.

Another glass case featured Hunter's work on bone growth and development. Prior to his experiments it was believed that bones grow uniformly—when they get bigger, it was thought, they get bigger all over. Hunter took young domestic fowl and bored two holes, at a carefully measured distance from each other, into their growing leg bones. He filled the holes with lead shot plugs. If bone growth occurs uniformly over the bone, then after a few months one should find the holes to be further apart from their originally measured distance. He discovered after many such experiments that the bones grew significantly larger (in their normal developmental pattern) but the distance between the two holes remained the same after months and years. This led him to understand that bone growth occurs by a process wherein new material is deposited on the external surface. New material is not added from within the bone structure, nor do bones grow uniformly over their entire area. Here, in the case before me, were a series of bird bones, cleaned and sliced through their length to expose the lead shot plugs. And next to them, illustrating the same point about bone growth, was a large liquid-filled jar containing a bright red pig's jaw that Hunter had stained by feeding the animal an alizarin dye for weeks before killing it.

Hunter dissected many criminals in his day, and he may have kept some resurrection men on a respectable salary as well. This morbid lifestyle made him rather unpopular with neighbors, and he subsequently equipped his home anatomy lab with strong shutters that could be bolted from the inside for security purposes.

Toward the center of the small museum is the enormous skeleton of Mr. Charles O'Brien, the "Irish Giant" (fig. 2.6). Hunter prepared this skeleton himself by boiling the corpse in a large kettle. In his *Essays and Observations* Hunter offers advice to other collectors on how to prepare bones for display and study. Boiling is, in fact, the most effective method, but if one cannot manage that procedure for some reason, Hunter says, one should "put the bones into a tub, with a loose cover, so as to let the flies get to them; they will fly-blow them immediately, and in a fortnight's time they will have entirely destroyed the flesh. However, this can only be done in the summer."

The fantastical curiosities throughout Hunter's small collection (and I haven't gotten to the pathology section yet) at first tend to distract the visitor from the overall organization of the cases. It took me

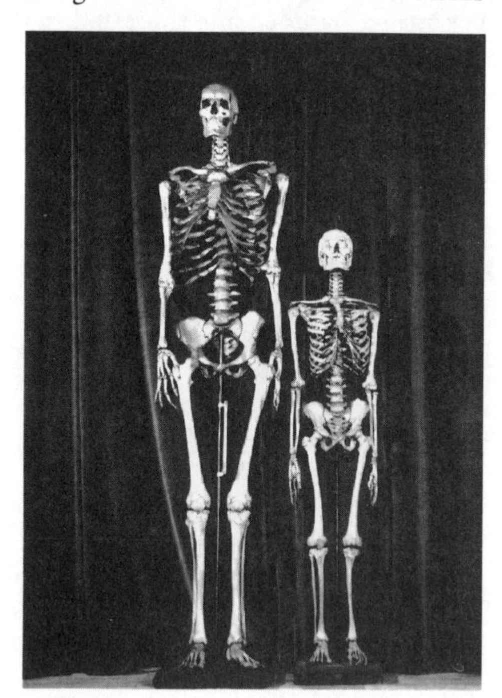

Figure 2.6. Skeleton of Charles O'Brien, next to a human skeleton of average size. (BY PERMISSION OF THE ROYAL COLLEGE OF SURGEONS, ENGLAND)

a while to pull back from the metaphorical trees and discover the forest of Hunter's plan.

Today Hunter's collection is arranged almost exactly the way that he set it up more than two hundred years ago. Walking through the museum is like walking through the past. But when students and curiosos of the nineteenth century studied the collection, the organization of the displays struck them as confusing. Matters were not helped when Hunter's manuscripts were destroyed in a fire shortly after his death. Some key to Hunter's display philosophy could have been found among his voluminous notebooks, but his brother-in-law Everard Home, to whom the manuscripts were entrusted, plagiarized the works mercilessly and then burned the originals. Hunter's friend William Clift, who was lobbying to have the British government purchase and protect the collection, once went to Everard Home's house to see about securing the manuscripts and was utterly shocked to find Hunter's brilliant manuscripts, a national treasure, being used by Everard as toilet paper. It's rather ironic that the Irish Giant's physical corpus received better posthumous treatment than Hunter's written corpus.

Despite the lack of Hunter's own writings, a great deal can be determined about his unique philosophy of specimen arrangement. One very influential naturalist, Richard Owen (who literally lived with the collection), eventually began to think differently about anatomical homologies (interpreting them as divine archetypes) after spending several years studying the collection. I fully understood the heretical nature of Hunter's presentation only after I had crossed the Channel and spent a few days at Georges Cuvier's Galerie d'anatomie comparée—but I get ahead of myself. The mystery of Hunter's collection would unfold slowly. For now, it was enough to realize that the large glass cases lining the walls were organized according to physiological system. As I examined the displays, I found radically different species crammed together in the same case. This ordering of things is not intuitively understood. Without a grasp of Hunter's underlying principles, the cases seemed slightly irrational. In one case, marked "Digestion," I found various dissections of mammals, parasitic worms, a cicada, a locust, slugs, a squid, a vulture, a woodpecker, and a puffin. This hodgepodge arrangement follows no known taxonomic grouping.

The thematic classification of the specimens falls into three major groups. When entering the museum, one finds the wall on the left to be populated with adaptation systems. Individual species, genera, and even families are displayed together to illustrate the unique ways that their structures fit the functions of digestion, circulation, respiration, and so forth. All these cases reflect the theme of individual survival. But Hunter did not respect traditional taxonomic divisions in his cases, and under the function of digestion, for example, he scandalously mixed vertebrates and invertebrates.

To the right of the entry a wall is filled with jars of reproductive organs and fetal development specimens. This second theme, of reproduction, is focused not around the individual organism but around the survival of the species. Here too, the different species are not contained in separate cases, but intermix with each other according to some higher unity of function. For example, the toad and the opossum are displayed together, violating every taxonomic classification, because they have both developed the habit of carrying their young on their backs.

The third major division in Hunter's collection comprises the cases at the far end of the museum space, separating the individual and species cases. These are the pathology cases, and they are almost, especially in their cumulative power, beyond description. Every conceivable way that nature can go wrong is collected here in frightening detail.

I was working down the cabinets on the right-hand side, meditating on a pileup of genitals (rhino, sparrow, mole, etc.) in a case marked "Reproduction," when I noticed a nightmarish specimen looming at the far end of the museum. When I flipped on the light switch for its case, I was greeted by jar after jar of floating calamity: hideously deformed human babies, pigs, and calves, and miscellaneous organs and tumors, all made more shocking by the inquisitive blade of Hunter's dissection. I was in the pathology wing of the museum (fig. 2.7).

The jar that first drew my attention was about the size of an industrial stew pot and contained a curdled mass of flesh. This menacing basketball-sized blob was a tumor that John Hunter surgically removed from a man's neck in 1785 (fig. 2.8). Next to the jar was a small card quoting Hunter's notes: "The operation was performed on Monday, October the 24th, 1785; it lasted twenty-five minutes, and the man did not cry out during the whole of the operation." This poor patient had a tumor, roughly the size of his own head, sprouting out of his neck, and Hunter cut it out of him sixty-odd years before anesthesia was discovered—with nothing to numb the pain except some swigs of whiskey. As I pondered many of the pathology jars, I wanted to get on my knees and thank the gods of experimental medicine for letting me be born in the twentieth century.

In the first half of the nineteenth century England's intelligentsia was dominated by the "argument from design." Natural theologians

Figure 2.7. A teratological "monster" fetus, from Hunter's collection. (BY PERMISSION OF THE ROYAL COLLEGE OF SURGEONS, ENGLAND)

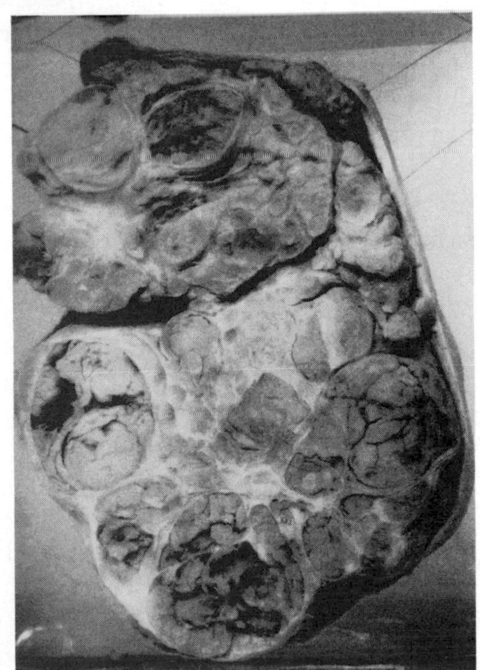

Figure 2.8. A tumor that John Hunter removed from a patient's neck in 1785. (BY PERMISSION OF THE ROYAL COLLEGE OF SURGEONS, ENGLAND)

were arguing that the natural world was perfectly adapted—each animal organ and appendage perfectly suited the peculiarities of different habitats and activities. Such perfect design, the argument concluded, proves the existence of a benevolent designer God. One of the overriding impressions that Hunter's pathology collection leaves on the observer, however, is that nature is sloppy. The notion of the perfect adaptation or fit of each animal to its environment and the elegantly coordinated physiological adaptation of each individual to itself (organs arranged and functioning in harmony) is dramatically challenged by Hunter's pathology jars. Shelf after shelf of biological "monsters" (the study of which was dubbed *teratology*, Latin for "the study of monsters," by the French anatomist Etienne Geoffroy Saint-Hilaire) gives one a glimpse into the imperfect workshop of nature. Indeed, Hunter's monster cabinets may have inspired some of Darwin's early meditations on evolution.

In the late 1830s and early 1840s Darwin was inadvertently spending quite a bit of time among Hunter's jars and simultaneously beginning his first diary notebooks on the subject of transmutation. The reason for his repeated exposure to the cabinets at the Royal College was his friendship with the Hunterian's curator, Richard Owen. Owen, amidst his herculean efforts to catalog Hunter's collection (for the approaching opening of the museum), was also studying the strange fossils that Darwin had brought back from South America. Owen and Darwin sat for hours and days among the monstrosities at the Royal College, hashing out the significance of Darwin's gigantic sloth fossils. Hunter's jars of "failed life" served as the backdrop for Darwin's and Owen's early discussions of extinction. Individual bodies and organs, preserved in the act of fluctuation from the norm, were suggestive of flux at the species level. In the diaries of that period Darwin shows himself to be much concerned with monsters, and long before hitting on natural selection, he puzzled over the role such monsters might play in transforming species. As I stood before the various jars of botched creations and maladaptations, I couldn't help but suspect that they had had some influence on Darwin's loss of faith in the perfect adaptations of a benevolent and omnipotent God.

Finally I came upon a display, in the center of the museum, entitled "Biological Preservation: An Historical Overview." Here, at long last, were the nuts and bolts of wet preservation techniques. Here were some answers to the mysteries of Peter the Great's decapitated specimens and Hunter's thousands of perfectly pickled organs. What intricate process, like the complexities of taxidermy, had gone into the maintenance of these ageless organic treasures? What elaborate precautions could sustain a man's stomach or a neck tumor for study by novice surgeons over the course of two centuries? What mystery fluid, bathing all the specimens, could bring a two-hundred-year-old fetus to my morbid gaze in the virtually unscathed condition of the day of its excision? The answer was humorously simple: booze.

I had been sure that some obscure alchemical secret kept the bottled organs in their vivacious state. But my philosopher's stone turned out to be liquor. Alcohol can kill you by pickling your liver, but then it can keep your liver fresh and immortal for centuries of scientific gawking.

Prior to 1666 the only manner of specimen preservation was the dry technique. Dry preservation is actually a family of related techniques, only one of which is taxidermy. In addition to the mounting techniques of taxidermy, many of the earliest museums practiced a method wherein the organ was injected with mercury, air-dried, and then heavily varnished. When I was a kid, I entered my school's science fair by displaying a series of preserved caterpillars. Unbeknownst to myself, I was actually practicing a crude form of dry preparation. I laid the dead caterpillars down on a hard surface, following the steps laid out in Vinson Brown's *Building Your Own Nature Museum*, and repeatedly rolled a pencil over the length of their bodies, forcing out all the innards through the anus.

When I had forced the corruptible juices out of the unlucky creatures, I used a lab bellows to pump hot air into the shriveled sacks that were their bodies. The result was a beautiful display of a caterpillar, but it was really only a dried-out, hardened shell (fig. 2.9). I was learning early on that scientific display is part education, part deception.

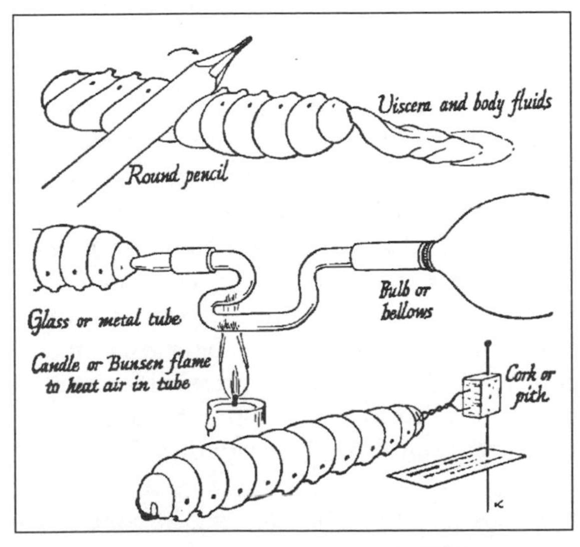

Figure 2.9. How to create a dried caterpillar specimen. (DRAWN BY DON GRAEME KELLEY IN BUILDING YOUR OWN NATURE MUSEUM, BY VINSON BROWN, ARCO PUBLISHING, INC. 1984. ORIG. PUB. 1954)

Some of the earliest curiosity cabinets, such as those of Oleus Worm (1588–1654) and Fredrick Ruysch (1638–1731), were filled with larger and more elaborate versions of my caterpillar method. Animals were relieved of their viscera, blown into three-dimensional balloons, and shellacked for display. According to Hunter's explanation of the dry prep procedure, varnishing the specimen has three benefits. First, the hardened oil-based coating will prevent the specimen from being destroyed by hungry insects. Second, "it preserves them from the dirt, as any dirt will more easily come off a smooth coat of varnish than off the preparation itself." And third, "it gives the preparation a much brighter colour, and makes the part shine by entering its pores, thereby rendering it more transparent." Many specimens were prepared in this way, but the majority of early organic specimens were simply mummified.

I grew up thinking that mummification was an erudite piece of human ingenuity—an elaborate technique in which mere mortals stopped the inevitable decay of the flesh. When I was quite young, I remember being dragged through the Field Museum's temporary exhibit of Tutankhamen's burial treasures. I didn't enjoy any of it at the time because most of the displays were too high for me to see. I knew that something important was bringing all this excitement to the Field, but I couldn't say what it was. Somehow in this confused experience, and in subsequent schooling about pyramids (together with late-night movies on TV featuring Boris Karloff), an association of images and concepts formed in my juvenile head. (The sad fact is that the real epistemology of how one's belief commitments get strung together has almost nothing to do with logic. Logic has the cleanup job of having to go in and sort through the chaos of our sense impressions. Actually, when I eventually interviewed museum designers and display developers—and there will be more on this in chapter 7—they said that these rather contingent associations of impressions were all they could hope for in putting together an exhibit. The most you can get at a modern museum is an association of sensual impressions, not the logical relation of concepts or the developmental relation of ideas. Developers and designers consider the exhibit successful if the impressions have been engaging. The issue of exactly what cognitive process occurs in museum education eventually began to occupy my attention after I spent time in the museums of London and Paris. When sensual impressions—sounds, images, tactile sensations—are linked together in our minds because they just happen to accompany each other in time, they frequently mislead us.)

It turns out that mummification is quite simple, an inevitable natural process that occurs to any corpse stored in a very dry place and protected from insects. Mummies are just completely dehydrated bodies, which, without water, become like bags of bones. The skin and muscle become dry and rigid, and the tissues generally shrink-wrap the skeletal armature as they lose moisture. Indeed, it was *natural* mummification, occurring in the desert climates south of the Mediterranean, that first led the Egyptians to formally reproduce the effect.

Very low humidity essentially eliminates maggot activity, and without these the body does not undergo the complete meltdown that ordinary decomposition entails. Organs still decay within the dehydrating corpse because natural cell death (autolysis) and putrefaction still occur, but the process is contained within the shriveling body rather than erupting out in its normal fashion. Full putrefaction is the process by which one's internal bacteria, ordinarily aiding in digestion during life, start to burst forth from the colon and consume the body, creating noxious gases and nightmarish bloating. Many naturally mummified bodies have escaped this process of becoming human soup only to become objects of veneration and wonder. Natural mummies have been found in the highlands of South America, in caves all over the world (including Mammoth Cave in Kentucky), and of course in Egypt.

The natural ability of organic material to dry out and stay preserved under the right climatic conditions was capitalized upon by early museum collectors. Mundane and exotic animals alike received the dehydration treatment and then found themselves lining the shelves of curiosity cabinets all over seventeenth-century Europe. Unfortunately, these air-dried, varnished mummies did not preserve their full-blooded living appearance. Shriveled and blackened versions of their former selves, these specimens were almost useless when it came to scientific study. Still, they made for good drama, and both human and animal mummies were used in morality plays (e.g., in the eighteenth century, the Brothers of St. Francis in Rome used to put on plays about purgatory using real dried corpses).

With the rising trend of empirical science in the eighteenth century, the increasing demand for specimens that were more lifelike led educationally minded collectors to create wax models. Brightly painted wax models of vivisected animals and dissected humans filled European collections in an attempt to render a more accurate representation of nature. The subtleties of the flesh, utterly lost in the mummified specimens, could be re-created with wax. When I eventually trekked to the Jardin des plantes in Paris, I discovered many poorly made wax models in Cuvier's Galerie d'anatomie comparée, but the gold mine of such creations is La Specola, in Florence.

La Specola is an amazing little museum, south of the Pitti Palace, that contains room after room of alarming wax figures. It is primarily a zoology collection of taxidermic and wax re-creations, but it also contains elaborate human dissection models. Grand Duke Pietro Leopoldo of Lorraine initiated the creation of these manikins in 1771 in order to teach human anatomy without the use of real human corpses. The modeled bodies are so disturbing because the artists, perhaps carried away by their love of detail, made the splayed and severed people look as though they are still alive. A reclining woman looks up at you as her intestines spill out in colorful detail. A skinless man seems to shriek and distort his face while he displays his lymphatic system to the patrons. Partly dissected heads stare wide-eyed as you pass one of the cabinets.

But, of course, here again the wax specimens, like the mummies, could not do justice to all the subtleties of organic objects. In Hunter's wet method, unlike the wax model technique, the actual organs could represent themselves—albeit abstracted from their context. And this liquid preservation, when done properly, was an infinitely more accurate form of representation than the shriveling dry technique. Strangely enough, the wet method was actually discovered almost a hundred years before Hunter began his exceptional practice.

It was Robert Boyle (1627–1691), the English polymath, who essentially discovered the wet technique in the 1660s. Boyle is perhaps best known because he has a law of nature named after him. Boyle's Law holds that at a fixed temperature, the pressure of a confined ideal gas varies inversely with its volume. We can also thank Boyle for naming the barometer. But in addition to his physico-chemical achievements, he was an all-rounder in every other field of natural science (then referred to as natural philosophy) and a prolific theologian as well. Boyle was part of the fascinating generation of English intellectuals who developed the Royal Society, and the founding of this society signaled a sea change in European thought.

Collections of artifacts and natural oddities had become all the rage in the century that followed the discovery of the New World. European gentry became childlike again in their appreciation of the marvelous for its own sake, and this spawned the age of the curiosity cabinet (Figs. 2.10, 2.11). Historian Wilma George points out, "For the most part, the cabinet of curiosities was just what it said it was: odds and ends to excite wonder. Almost every collection had 'monsters' in it: 'a monstrous calf with two heads,' 'a horned horse' and an 'ovum magicum,' 'a calf with five feet' and 'ova monstrosa.' When the collection was displayed it rarely had any order in it. Stuffed

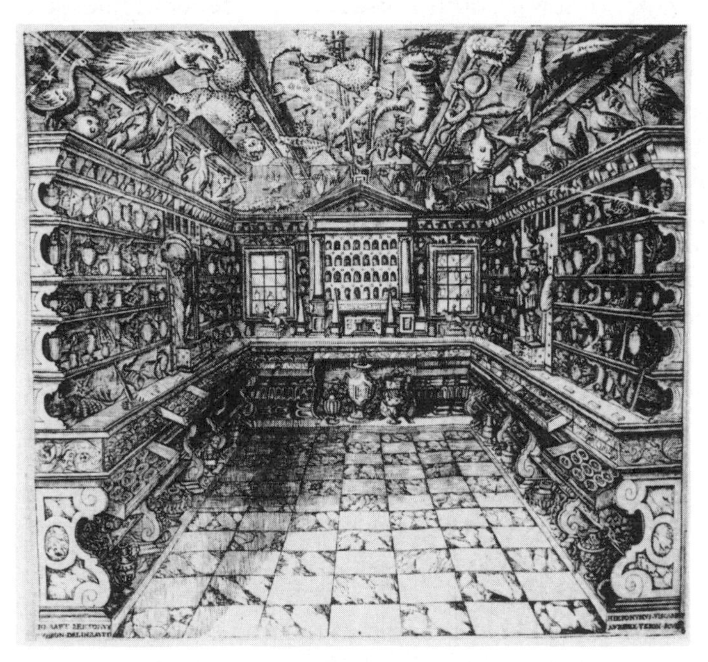

Figure 2.10. The museum of Francesco Calceolari in Verona, 1620s. A typical pell-mell curiosity cabinet of the early modern age. (THE ORIGINS OF MUSEUMS, EDS. O. IMPEY AND A. MACGREGOR, 1985, BY PERMISSION OF OXFORD UNIVERSITY PRESS)

Figure 2.11. The Copenhagen museum of Oleus Worm (1588–1654), from *Museum Wormianum*, 1655. (THE ORIGINS OF MUSEUMS, EDS. O. IMPEY AND A. MACGREGOR, 1985, BY PERMISSION OF OXFORD UNIVERSITY PRESS)

birds stood on top [of] the shelves of Calceolari's museum because there was room—mammals and fish hung from the rafters."

These precursors to the modern museum were generally housed in private homes, inaccessible to the public and primarily designed to engross the owner and his society friends. Education, about the principles or the causes that undergird nature, was not the function or interest of the curiosity cabinet.

In 1627 a short work by Francis Bacon, entitled *The New Atlantis*, was posthumously published. In this brief piece of fiction Bacon (1561–1626) told the story of a ship of explorers who get lost in the "South Sea" and eventually stumble upon a previously undiscovered but highly advanced island community. These technologically superior people, the "Bensalemites," demonstrate their political, economical, ethical, and intellectual superiority to the stupefied crew throughout the story, and the narrative culminates in a trip to "Solomon's House."

Solomon's House is a fantasy that Bacon invented to embody, in literary form, the growing scientific ideals of the seventeenth century. The crew members are allowed to see the amazing experiments and the technology of this utopian science institute. The high priest of Solomon's House explains the mission of the project: "The end of our foundation is the knowledge of Causes, and secret motions of things; and the enlarging of the bounds of Human Empire, to the effecting of all things possible."

In a classic case of life imitating art, Bacon's fictional Solomon's House became the chief inspiration behind the creation of the real Royal Society. Men of learning wanted to create an institution where they could come together and pool their wisdom, feeding off each other's experimental data and thereby advancing the human condition. To that end, they embraced the philosophy of empiricism, and they started to create a collective repository of knowledge for education and research.

The Royal Society was an enormous consolidating force that sought to bring the disparate collections of curiosities together. As empiricists, the members believed that patterns in nature can be discovered only through the collection of extensive data; they cannot be spun de novo from the minds of armchair philosophers. With this guiding principle the Royal Society placed an announcement in the October 1666 edition of its publication *Philosophical Transactions*, encouraging people to donate their private cabinets to the society's future museum. The idea here—and it was catching on all over the West—was that the aesthetically pleasing objects of the separate curiosity cabinets could reveal deep truths about causes once they were brought together and studied comparatively and analytically. Understanding causes, the logic went, leads to greater manipulation of nature, and in this manipulation lies the secret of humanity's progress. The curiosity cabinets were being consumed and transformed by the scientific revolution. The act of hoarding things and displaying them in groups was changing its function.

Robert Boyle was a principal player in this revolution not only because he contributed his perceptive observational skills to the cause but, more important, because he helped provide a philosophical foundation for the changing function of collections. Along with other natural philosophers such as Robert Hooke and René Descartes, Robert Boyle preached the doctrine of the microworld. Prior to the Royal Society movement, collectors hoarded fascinating forms such as peacock feathers, shark teeth, sea sponges, and so forth. Boyle and other scientists now had a system for explaining the causes of these interesting forms. Boyle even wrote a book called *The Origin of Forms and Qualities* in which he systematically revealed the hitherto unknown microstructure of everyday objects. Using the very recently discovered microscope, men such as Boyle and Hooke described the tiny microcosmos that caused our macrocosmos to take its particular shape and form.

Again, it takes an imaginative leap for us to think our way back to an age when the microscopic world was just being discovered. How exciting and intoxicating it must have been to discover cells in the leaves of plants, sediment particles hidden in fluids, and the structural geometry of minerals and crystals. It is small wonder that Boyle and others saw themselves as holding the key to all the marvelous diversity of the macroworld. Very different material things, they argued, were really made up of the same stuff, but this stuff (atomic particles) had different shapes and sizes in the microworld, and these structural differences were reflected in the big world of peacock feathers, shark teeth, and sea sponges. As the Royal Society's *Philosophical Transactions* states in 1666, the microparticles (recently discovered by the microscope) reveal "the subtlety of the composition of bodies, the structure of their parts, the various texture of their matter, the instruments and manner of their inward motions, and all the other appearances of things . . . whence may emerge many admirable advantages toward the enlargement of the active and mechanic part of knowledge, because we may perhaps be enabled to discern the secret workings of Nature." Boyle was finally paying on the promissory note that Bacon had fantasized in *The New Atlantis*—the curiosities of nature were becoming subsumed under the umbrella of physico-chemical laws.

Apart from his philosophy of the microworld, of course, Boyle was a quintessential laboratory man. In May 1666 Boyle stood before the other fellows of the Royal Society in London and shared yet another of his striking lab discoveries. He revealed a number of jars that contained a progressive series of fetal developments in chickens. Here was a day-by-day parade of chick fetuses, each arrested in a different stage of development—and each in its own jar. But it was not the science of the embryological development that Boyle was excitedly sharing with his colleagues. This time it wasn't the underlying microcauses that he was explaining. It was his discovery that the "spirit of wine," as he called it, was the very best embalming fluid that anyone had seen to date.

One might be slightly unimpressed about Boyle's "brilliant" discovery of dropping specimens in booze. It turns out, however, that it wasn't just wine that Boyle was employing as preservative—it was "spirit of wine." And before you could pickle things in spirits, the spirits had to exist.

"Spirit of wine" was the term used to describe the more concentrated ethyl alcohol that could be distilled from wine. But the world was generally unfamiliar with this distillation technique until the sixteenth and seventeenth centuries, so Boyle's discovery is, in fact, not so late as one might think. Fermenting rice, oats, barley, grapes, and so forth into beer and wine was, of course, a very old technology in the East and West. But distilling wine and beer into harder alcohol was not possible until alchemists understood that alcohol and water have different boiling points.

Water boils at a temperature of 100°C (212°F), and ethyl alcohol boils at only 78.3°C (173°F). This difference makes it possible to boil out the alcohol from a beverage such as wine while leaving the water and other substances behind. This distillation is accomplished by heating the wine to a temperature above 78.3°C but below 100°C. The vapor is captured and condensed into a liquid of considerably higher alcoholic concentration. Repeating this process over and over was how Boyle created the potent "spirit of wine." Boyle describes his pickling technique to the Royal Society:

> I usually found it convenient to let the little animals, I meant to embalm, lie for a little while in ordinary spirit of wine, to wash off the looser filth that is wont to adhere to the chick when taken out of the egg; and then, having put either the same kind of spirit, or better upon the same bird, I suffered it to soak some hours (perhaps some days) therein, that the liquor, having drawn as it were what tincture it could, the fetus being removed

into more pure and well dephlegmed spirit of wine, might not discolor it, but leave it almost as limpid as before it was put in.

Boyle discovered the wet preservation technique, but it wasn't until John Hunter that the process was perfected to the level of art. In order for the specimens to maintain the "limpid" quality that Boyle describes, their containers had to be sealed in a more rigorous fashion. All of Hunter's preserving jars were sealed first with a cap of pig's bladder, then with a cap of tin, and over that a seal of lead, and lastly another pig's bladder stretched over the top. The nearly imperceptible support filaments that suspended the specimens were linen threads soaked in molten beeswax.

It turns out that the linen thread technique is still the preferred method at the beginning of the twenty-first century, but the alcohol solution has finally given way to the use of formaldehyde. The advantages of formaldehyde over Hunter's alcohol is that less shrinkage and bleaching occur in the modern solution. In addition to this change in preserving liquid, the advent of plastics in the 1940s proved to be an advance over the glass jars because acrylic containers can be custom-built for each specimen. These customized containers cut down on the visual distortions of round glass jars.

From my excursions I had gained considerable insight into the how of eighteenth- and nineteenth-century museum display, and now the more difficult questions of why began to emerge again. As I studied the tall mahogany cabinets surrounding me, I understood that Hunter's collection was different from his predecessors' not only because of the wet technique. It was laid out according to underlying philosophical and scientific principles, whereas the earlier curiosity cabinets had been arranged according to aesthetic and moral principles.

I began to notice some of Hunter's motivations in the displays. For example, the circulation cabinets began with animals that had very rudimentary circulation systems, then proceeded to animals with more distinct systems, and continued through greater and greater complexity, finally culminating in the human heart and circulation system. Each of the other cases seemed to follow this basic sequence of increasingly complex forms associated with given physiological functions. As I stared at the collection, my mind could not help but appreciate the unity and continuity of all life. And I discovered later that this feeling of unity and continuity was precisely what frightened many scientists who viewed Hunter's unique displays.

Feeling a little burned out on museology for that day, Heidi and I sat on the green leather couch and discussed our options for dinner. As we were contemplating the virtues of a fresh pint of Guinness stout and resolving to make our way to a nearby pub, a middle-aged woman and a young boy entered the museum. As he entered, the boy, approximately ten years old, seemed reluctant to waste the sunny day in this obviously boring place designed for grown-ups. But the woman, who I assume was the boy's mother, began to flip on the light switches of the bulging and busy cases. His voice slowly began to rise, punctuating the quiet room with gasps of amazement.

"Oh, my God!" he repeatedly shrieked as he moved through the reproduction cases. I had wanted to cry out in the same way earlier, but the proprieties of adulthood squelch one's expressions of amazement and wonder.

"Oh, Lord, I can't believe it!" he gasped in his thick northern English accent. He was moving into the pathology section of the museum now, and he was being drawn into the morose magic of Hunter's collection. As he kneeled in front of a fetus with two fused heads, peering at it intensely, his mother suddenly turned to him and asked, "Is this disturbing to you, William?"

He didn't look away from the cases, but responded, "God, yes, very."

"Shall we go, then, dear?"

"No," he shot back, "absolutely not."

Heidi and I chuckled at this illuminating mother-son exchange as we left the museum. I couldn't help wondering whether mothers and sons had blurted out similar exchanges over two hundred years ago when they stood in front of Foma and the nearby jars of ex-lovers' heads.

Of course, unique specimens in isolation, even fantastic ones, are not scientifically illuminating. During Hunter's lifetime curators began to understand that the intellectual benefits of collecting and displaying were created by the theoretical arrangement or organization of the items, not just by the sight of the items themselves. To understand this aspect of museology, we need to explore the science of arranging and classifying, otherwise known as taxonomy.

CHAPTER 3

TAXONOMIC INTOXICATION, PART I: VISUALIZING THE INVISIBLE

PERHAPS THE MOST NOTABLE ASPECT of the Hunterian Museum—besides the bleached and bloated "monsters"— is that it's not intended for a lay audience. The Hunterian doesn't really care whether the average person is "getting it." This is unexpectedly unnerving, since most other museum encounters come complete with torrents of helpful information. Almost every museum specimen that you have ever encountered, be it a fossil, a jar, or a mount, is accompanied by a plaque, a chart, or a recording that announces to you—even before a question can be formulated in your head—"What you are seeing is thus and so. . . ." One's perception is immediately linked up with some conceptual content; in fact, many people (usually moving too slowly for their spouse's taste) will look at the next diorama in a museum only after they have read the written material. I occasionally adopt this tactic in a new museum, because I want to be sure that I comprehend what I'm seeing. It is extremely easy to look long and hard at something but still fail to take in its significance. When Dr. Watson witnesses the same events as Sherlock Holmes and expresses frustration at Holmes's ability to read off the solution from these events, Holmes retorts: "You *see*, Watson, but you do not *observe*." In other words, perception alone, without intellectual inferences and connections, is relatively blind. And while some of the Hunterian surgical specimens were duly explained, I felt like a rather myopic Watson when I stared into his natural history armoires. Unlike public museums of natural history, such as the Field Museum or the American Museum of Natural History, the Hunterian's cabinets seemed obdurate and insensitive to the visitor's personal quandaries. There is an interesting tension between the collection's profligate and extravagant oddities, on one hand, and its stubborn theoretical silence, on the other.

What was Hunter trying to communicate when he grouped his specimens? This question alone gives one a more interactive relationship with the museum than any computer gadgetry could. Some system of organization is emanating from the mute specimens, an emerging cognitive order, but one is hard pressed to articulate it. After all, even the chaotic piling of creatures that filled the early curiosity cabinets had a purpose, an underlying but persistent agenda: to show that God is prolific, prodigious, and ingenious. At the end of the Renaissance and in the days before the Royal Society, a cabinet's qualitative variations and quantitative excesses themselves heralded a message to the spectator. Something invisible—a feeling or cognition of the Deity's fertile presence—was being conveyed in the visuals of the cabinet. Konrad Gesner, for example, explains to his sixteenth-century patrons that even his collection of lowly insects indicates God's power: "These little creatures so hateful to all men, are not yet to be despised, since they are created of Almighty God for diverse and sundry uses. First of all, by these we are forewarned of the near approaches of foul weather and storms; secondly, they yield medicines for us when we are sick, and are food for diverse other creatures, as well as birds and fishes. They show and set forth the Omnipotency of God, and execute his justice."

In the late eighteenth and early nineteenth centuries naturalists such as Hunter in London and Cuvier in Paris wanted to convey a different sort of agenda through their cabinets. It was not the abundance and generosity of nature's creativity that they sought to celebrate, but the rationality and orderliness beneath the profusion and confusion of forms. As curators, Hunter and Cuvier wished to convey new ideas, to visualize newly discovered invisibles. But to say that naturalists sought to display systems is trivial. What needs knowing is which systems they wanted to display and why they wished to display them. In order to understand the way in which collections communicate theoretical and ideological commitments, it's imperative that we tour the history of taxonomy. The next two chapters will trace the paradigm shifts in European taxonomy and explore the ways in which these had an impact upon museology.

From my researches into Hunter and his collection, I was led to Baron Georges Cuvier (1769–1832), who developed the only European museum to hold a candle to Hunter's. As in the case of Hunter's collection, Cuvier's original cabinet, begun around 1802, was partially preserved in its original format. The current Cuvier museum is a little window into the history of museology. I resolved to make a pilgrimage to Paris, to Cuvier's collection, housed somewhere in the Jardin des plantes. The fact that Cuvier's displays were so temporally proximate to Hunter's and that Cuvier had visited Hunter's museum on more than one occasion led me to think that this trip could unlock some mysteries.

Once in Paris I discovered that Cuvier's collection at the Galerie d'anatomie comparée was not open on Tuesdays. So I thought to waste no time and instead visited an obscure little museum called the Musée de la chasse et de la nature. You won't find this one in the tour guides. Officially dedicated to the "celebration of hunting and those who love and preserve nature," it seems paradoxical: a museum for those who both love nature and enjoy killing animals from time to time. It is located in the historic Hotel Guénégaud (built in the 1640s) in the Marais, in the fourth arrondissement. The beautiful complex of buildings became a hunting museum in the late 1960s. When I arrived, the entire staff was out to lunch. Then around two o'clock the curators and staff returned, almost surprised to find me interested in paying the admission.

The first few exhibits I encountered are dedicated to the history of hunting weapons. Actually, the word *history* suggests more educational content than is in fact present. Most of the cases are simply aesthetic arrangements, artistic placements of objects rather than historical sequences. For example, one case contained crossbows, rifles from various periods, and gunpowder flasks. The only organizing principle here seems to be that they all share intricate ivory inlay work.

The museum path proceeds into an Africa hall and an Americas hall, where various stuffed animals of those regions are hung on the walls and posed in the corners (fig. 3.1). The taxidermy is excellent. The displays, laid out as *haute art* in lavish baroque rooms, were designed for the edification of committed (and wealthy) hunters. There were huge gaudy paintings celebrating the hunt, including a romantic canvas depicting a crimson fury of hunting dogs converging on a warthog. Upon closer inspection, an elegant chandelier dangling above me revealed four animal hoofs that had been formed into decorative highlights. There was no attempt to educate the patron on the history of hunting methods, and there was definitely no attempt to offer alternative (or dissenting) views of hunting, only an inspirational tribute to the inherent nobility of it all.

I could hear some posh-sounding banquet occurring in a closed-off branch of the museum. I was the only museum patron in the collection halls, so I turned to the security guard (who had been following me around the entire time—out of boredom rather than suspicion) and queried him about the gustatory din. The museum, he explained, housed an active hunting club, or more accurately, the

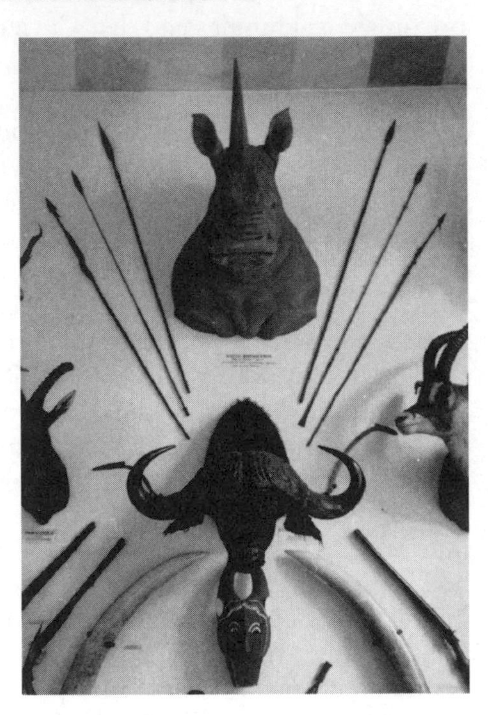

Figure 3.1. A wall display of taxidermy mounts in the Musée de la chasse et de la nature, Paris.

active hunting club housed a museum. The club was lunching that afternoon. This was a museum by and for an elite group, and I began to suspect that visitors were being tolerated in order that some funds from the Ministry of Cultural Affairs could continue to flow. I asked if I could walk in the manicured garden in the courtyard, but the answer was a firm no.

The relationship between any museum and its funding is complex. Generally speaking, tracking the flow of money (public or private) provides many explanations of why curators curate the way they do, and even why one particular curator gets the job in the first place. When a government is a monarchy, as it was during the invention of the Jardin des plantes (originally a royal garden of medicinal herbs, then a royal botanical garden), academics tend to resent any constraint on their intellectual freedom and keenly feel the watchful eye of king and court. Georges-Louis Leclerc, comte de Buffon (1707–1788), the director of the royal botanical garden, for example, had to answer to Versailles for everything, from the borrowing and returning of scientific instruments to the hiring and firing of colleagues and the purchasing of new and rare collections. But the French court also had the commendable reputation of moving beyond its own outdated science institutions by opening new ones. As Jacques Roger points out, "Louis the XIII transformed the Royal

Lectures into the Royal College and, inspired by Richelieu, also created the Royal Botanical Garden. Thus began a tradition of French administration, which still endures: every time the government deems that the university has refused to adapt itself to the requirements of new scientific theories, it creates a new institution." Counterintuitive though it may seem, the government can actually be more progressive than the scientists.

The issue of who financially supports science institutions and their collections was a point of national pride during the time of Hunter and Cuvier. The members of the Royal Society in England were unrepentant braggarts about their superior private status. Unlike the French institute, they received no money from the government. The Royal Society's historian, Charles Weld, echoes the society's sense of superiority when he states, "It would be repugnant to the feelings of Englishmen to submit to the regulations of the [French] Institute, which require official addresses, and the names of candidates for admission into their body, to be approved by Government, before the first are delivered, or the second elected."

Submitting science to governmental funding and control means that only those research programs that politics deems to be useful will be properly funded. But following up on seemingly useless topics is, according to the original hoarding philosophy of the Royal Society (remember Bacon), the very essence of good science. Thomas Malthus (1766–1834) speaks of the need for government, and everybody else, to leave science alone:

> If science be manifestly incomplete, and yet of the highest importance, it would surely be most unwise to restrain inquiry, conducted on just principles, even when the immediate practical utility of it was not visible. In mathematics, chemistry, and every branch of natural philosophy, how many are the inquiries necessary for their improvement and completion, which, taken separately, do not appear to lead to any specifically advantageous purpose; how many useful inventions, and how much valuable and improving knowledge would have been lost, if a rational curiosity and a love of information had not generally been allowed to be a sufficient motive for the search after truth.

This plea for autonomy is impressive, even vaguely inspirational, but it also sounds a little quaint and unrealistic in our age of Hiroshima and cloned sheep.

In the twenty-first century it is nigh impossible for public science museums, especially the large-scale freestanding institutions, to survive without major governmental support. Large institutions that originated around the turn of the century, such as Chicago's Field Museum, Los Angeles's Natural History Museum, and New York's American Museum of Natural History, currently require between $50 million and $100 million for physical renewal. Museum foundation support, admissions, and corporate and individual donations go a long way toward sustaining annual operating costs. But without government grants, physical plant renewal will be unlikely, and without increased partnerships with major corporations (e.g., the Field Museum's recent partnerships with McDonald's and Disney World), today's museums will not be able to compete for audiences. A growing trend in England is to revitalize older museums by using government lottery funds. The contemporary problem of funding for large-scale natural history museums may eventually force collections to become divided up, parceled out, and privatized again during the next century.

When I did go to the Jardin des plantes, I entered through the north side, winding through the curving paths of a lush but neatly trimmed forest. This vaguely wild grove lets out into a brilliant expanse of perfectly geometric flower beds. The enormous grid of flora is framed on either side by wide promenades with canopylike tunnels of tree cover. Everything is living, but the garden is strangely inorganic in its crystalline precision.

Around to the Seine side of the Jardin is the neoclassical facade of the Galerie d'anatomie comparée. The entrance foyer is very dark, especially after the intensity of the garden colors. It smells like an old church. Once again, Heidi and I were the only ones there.

Out of the grottolike darkness emerged a bizarre image to our right. As my dilating pupils struggled to meet the challenge of this new environment, I began to make out a naked man writhing on his back. He was being attacked by some creature. Even before we moved closer, the stillness of the figures quickly betrayed their marble composition, but what a composition! The statue is of a young man, naked except for the telltale bone-and-feather accessories of the native, being throttled by a demonic-looking orangutan while a baby orang joins in with murderous zeal. The sculptural rendering is extremely realistic. The victim's face is horrifically stressed, with mouth wide open; it is clear he is in his last conscious moments before being killed dead. The orang's strangling arms are taut, earnest choking vises, and the ape's visage is nightmarish. The orang's face is a strange paradox of twisted mouth and doughy cheek, with eyes sunken beneath the bulbous brow. The violence was lovingly rendered, according to a small bronze label, by a sculptor named Emmanuel Fremiet (1824–1910).

It was hard to interpret the point of the sculpture, and the plaque listing the title, "Ourang-outang etranglant un sauvage de Borneo (1898)," was not very helpful. Was the Borneo "savage" about to kill the baby orang when the parent stepped in to defend? Was this a noble representation of the creature? Or was the young man minding his own business when the odious creature indulged its natural aggressive tendencies on the innocent victim? It is an ominous entry sculpture, a kind of brutal antidote to the perfectly manicured nature of the gardens outside.

Beyond the foyer, in the hangar-sized wooden hall, I actually startled at the sight of the main room. Here are hundreds of skeletons seemingly charging toward the visitor. Right in the center is a human model stripped of skin, arms outstretched, poetically exposing muscles (fig leaf in place). Man is at the center of this ossified stampede (fig. 3.2).

Before we can make sense of what is laid out in Cuvier's gallery, it is important to excavate some of the history of classification. And before we dig in this direction, it is of paramount importance to examine *why* we classify things at all.

Of course, from the philosophical perspective, a little reflection demonstrates that it is impossible *not* to classify. To state a proposition or to think a series of coherent thoughts is to categorize in some way. Language and thought relentlessly follow some basic rules of logic (despite the protests of romantics, who undo their

Figure 3.2. Human *écorché* (complete with fig leaf), standing amidst the skeletal "stampede" in the Galerie d'anatomie comparée, Paris. (PHOTOGRAPH © RICHARD ROSS)

objections by speaking), and logic is nothing but a formalized system of categorization.

Take the very straightforward case of the law of noncontradiction. This law states, "A or not-A." Either you're pregnant or you're not; it cannot be both. One cannot actually think a contradiction; that is to say, one cannot hold a real contradiction before the mind's eye. When I ask my students if it is truly possible to be a married bachelor, they will do linguistic back flips attempting to demonstrate its possibility. We might change the meanings of the terms as a sleight-of-hand way around the challenge, or spread the proposition over time (as in "He was married last year but is now divorced"), but the rule remains doggedly resolute: If you were once married but are now divorced, then you are not married and hence there is no real contradiction. Someone will offer: "What about if you are married but behave like a bachelor?" But all one has done here is change the meaning of *bachelor* from "unmarried man" to "hip two-timer." It seems like a contradiction, but it is only a word game. You cannot be a married unmarried man. Contradictions are incoherent meltdowns; they are radical category violations.

To have a concept, any concept, is to have its negation already in tow. If you think of a thing—say, a dog—you already have two cognitive categories: "dog" and "not-dog." There is a class of things called "dog," and there is a class of things (quite substantial, in fact) that are "not-dog." Childishly profound as this is, it is actually relevant. Language and thought cannot really function without this most basic tool for carving up reality. Human beings seem to assign objects to conceptual boxes (object X is a dog, so it goes into the conceptual box "dog") by checking a short list of properties. Using the law of noncontradiction, one asks: "Does X have (or not have) the property 'four-legged'? Does X have (or not have) the property 'canine teeth'?" Likewise, how do we know if something should be categorized as a circle or not? Same process. Is X (or is it not) a planar figure? Is it the case (or is it not the case) that all points are equidistant from the center? And so on.

Anthropologists, psychologists, and philosophers have long puzzled about how the human mind creates this conceptual filing cabinet of the world. In fact, Emile Durkheim (1858–1917), one of the patriarchs of modern sociology, argued that this most primitive categorization system (A or not-A) must have originated in our animal prehistory as a version of "us or them." Developing categories by which protohumans could cognitively organize their experience must have emerged directly out of social interaction. Durkheim speculated that the first significant repetition of the "not" category was when a clan of creatures perceived another clan as "not us," and the concept "or" arose, as in "Either you're with us *or* you're against us."

As children, we learn to categorize the world very early, but it is still a process of learning and discovery—we may be born with taxonomic ability, but we are not born with a specific taxonomic system in place. Children learn to put things in category boxes through a variety of pressures: cultures serve up their time-worn categories, physiological proclivities make their preferences known early, and reality just seems to come in certain undeniable chunks. Ask children of different ages this question, and you can actually witness the developmental path of categorization: "Does the number 2 smell sweet?" At a certain age, kids will just laugh at this because they recognize the category mistake.

However humans actually begin the process of categorizing the world, it is clear that we are masters of the game. This is really no surprise, since survival comes more easily to those who can discriminate food from poison and friend from enemy. These survival aspects of taxonomy are fairly nonnegotiable. For example, certain kinds of plants kill you and certain kinds of plants heal you, and the very earliest observers of nature needed some way to recognize classes of plants and communicate the salient features to others. If your culture has no taxonomic distinction between hemlock and parsley, you're in trouble. Reality is stubborn, and it is in a culture's best interest to get its taxonomic system to conform to it, rather than waiting and hoping for the reverse to happen. But apart from these dramatic life-and-death classification scenarios, the vast majority of the ways in which we carve up nature involve rather arbitrary, though culture-bound, choices.

The word *taxonomy* is derived from two Greek words: *taxis*, meaning "arrangement," and *nomos*, meaning "law." And while we eventually want to focus on the biological traditions of taxonomy, we must recognize that the lawful arrangement of things by human beings is an older and more fundamental practice than the science of biology itself. Jorge Luis Borges, in his piece "The Analytical Language of John Wilkins," tells of a humorous taxonomic system that carves up the world in a very peculiar way. Borges cites a fictional Chinese encyclopedia entitled *Celestial Emporium of Benevolent Knowledge*, in which it is written that animals are divided into the following categories:

(a) belonging to the Emperor
(b) embalmed
(c) tame

(d) suck[l]ing pigs
(e) sirens
(f) fabulous
(g) stray dogs
(h) included in the present classification
(i) frenzied
(j) innumerable
(k) drawn with a very fine camel-hair brush
(l) et cetera
(m) having just broken the water pitcher
(n) that from a long way off look like flies

Michel Foucault reiterates this passage and claims that as he was laughing at the beautiful absurdity of this exotic system, it crept into his mind that he could not be sure that his own taxonomy of the world was not equally skewed. Perhaps you and I are so accustomed to our categories that they seem utterly natural, certain, and beyond question. But are they?

To follow the development of modern museum collecting is to follow the evolution of European classification, and we must follow this in order to understand three things. First, we want to explore the relativity of taxonomic systems, that is, how they change over time and adjust to social and intellectual developments. Second, we want to understand how Hunter's and Cuvier's curatorial practices fit into this larger story of classification. And last, we want to see how these same issues, of theoretical taxonomic dispute and practical museology, play themselves out in our post-Darwinian era.

Before we get into early taxonomy, let me give a brief example from anthropology that illustrates two important points. I give the following example of racial theory to illustrate how the seemingly dry science of classification can go to the very heart of human drama, and to illustrate the fact that museum display practices can communicate theory (and ideology) very effectively.

Unless one opens a box and dumps the contents on the table, displaying things involves automatically grouping them in ways that signify (and, as we suggested earlier, even this dumping approach may be signaling a theological theme—not to mention the question of just why these things went into the dumped box in the first place). Animals, plants, and even people have been categorized differently over the last five centuries, and museum displays have sought to create, reinforce, or challenge these ordering frameworks. A classic example of this dialectic between ideology and exhibition practice is the eighteenth- and nineteenth-century race debates. Briefly examining this more obvious anthropological example of "visualizing the invisible" will help render the subtler natural history cases more distinct.

Natural philosophers (this was the term for scientists, before the term *scientist* existed), including people such as Cuvier, Thomas Jefferson, and J.F. Blumenbach, debated with great seriousness the question of whether the human races had one origin (the theory of monogeny) or many origins (polygeny, a theory so popular in the United States that it was actually called the "American school of anthropology"). Was there one Adam and Eve that gave rise to all the diverse races, or did each race have its own Adam and Eve? This debate had major political and ethical implications (which, unsurprisingly, heated up during the abolition/antiabolition forensic), and it was essentially a debate about classification. If Negroes and Indians had the same great-great-great-great-great-great-etc.-grandparents as European Caucasoids, then the Europeans would have to acknowledge that "they" and "us" are in the same conceptual box. This proved difficult to acknowledge, since it implied, for the European, that he should take his boot off his brother's neck and help him up. Helping Europeans feel more comfortable with their hegemony was the alternative theory of races, polygeny. Here the idea was that each race bore the stamp of its separate creation, and the Asian, African, European, and Indian populations were more like different species rather than different races. In other words, "they" are not really "us." And if you don't get classified with the dominant species, then you're in for hard times. If it can be established that the races are unbridgeably separated, then it becomes easier to claim that some races are higher or lower than others.

Obviously, I've simplified this a bit, and we have to recognize through the webs of politics and prejudice that there was a real mystery and valid scientific puzzle about how and why people have different physical traits. Many natural philosophers were simply working on this question in an honest way (thinking about internal and environmental causes, examining adaptive possibilities and limitations, and so forth), and their theories were co-opted and abused by politicians and ideologues.

When Louis Agassiz (1807–1873), a former student of Georges Cuvier, arranged the Museum of Comparative Anatomy at Harvard, he sought to "educate" patrons in, among other things, his anthropological theories. It is a testament to the power and influence of John Hunter's and Cuvier's collections that when Agassiz set about creating the Harvard museum in the late 1850s he explicitly hoped to rival those European institutions. He stated that his chief objective was to construct an American museum that "would be as important for science as those founded by John Hunter in London or by Cuvier in the Jardin des plantes."

When Agassiz displayed the various races of man, he did so along the polygenic principles of his teacher. Agassiz believed that all distinct kinds of organisms, including traits such as skin color, owed their origins to special creative acts by God. For Agassiz, the Negro, the Caucasian, the Hottentot, the Mongol, and others were created to fit into different environments, and there was no such thing as evolution from one race to another. His principles (or his fears) also led him to argue that miscegenation led to sterile or enfeebled offspring. The patron who toured the Harvard museum would find a display that authoritatively reinforced the idea of separate creations. Dioramas and displays emphasized the discontinuity rather than the continuity of people. Race specimens were exhibited in their own isolated habitats.

While Agassiz built his museological arguments on a creationist (antievolutionary) premise, his own students became the generation to embrace Darwin's ideas. With regard to museum exhibits and the race question, this embrace of evolution meant that some things changed, but others remained the same. The World's Columbian Exposition of 1893 (where the Field Museum was born) was a six-month fair that transformed Chicago's lakefront into the utopian fairgrounds-cum-village called the "White City." The supervisor of the exposition, G. Brown Goode (a student of Louis Agassiz), understood that museum and fair exhibitions had a huge educational potential for the more visually oriented working classes. The central idea that he wanted to convey in all the exhibits was the idea of progress (a very popular ideological theme in Darwin's century). Goode stated that "the people's museum should be much more than a house full of specimens in glass cases. It should be a house full of ideas, arranged with the strictest attention to system." And when this philosophy was applied to the exhibition of the peoples of the world, the system became one of racial hierarchy.

Frederic Ward Putnam (1839–1915), a Harvard naturalist and another former student of Louis Agassiz, was brought to Chicago to set up the anthropology exhibits. While Putnam was not a polygenist, his ordering and categorizing of cultures was still very insulting to nonwhite races. Yes, there may have been some distant evolutionary fraternity between the races, as Darwinism implied, but the anthropologists of the time argued that some races had evolved more than others. So, using a combination of insulting cultural habitats and a color-coded placement of racial representatives, Putnam displayed an ascending chain, from "backward" dark races to "civilized" light races. He devised little villages of ethnic life (rife with stereotypes) that effectively reinforced Americans' belief in their economic, cultural, and possibly even biological superiority. The lesson that the museum gave to people of European descent was that they were winning the progress competition, and the lesson that it gave to non-Europeans was the obvious correlate.

Anthropology exhibits during the nineteenth century nicely illustrate the fact that specimen classification and arrangement can be rhetorical, can be persuasive. Let us widen the focus beyond the process of theoretically constructing humans (anthropology) to the process of constructing nature itself; and let us begin with the premodern taxonomy of nature.

Just as the Chicago exposition's patrons could read the hidden text of these anthropology exhibits, so too the early natural history displays were curated to be read in certain ways. In fact, it could be argued that in the early days, before the Royal Society began, the collections were almost exclusively about hidden text. In addition to the curiosity cabinets, which sought to manifest God's fecundity and elicit awe, there were moralizing and occult systems of organization.

The medieval period was permeated by a rather symbolic understanding of nature, and this perspective remained vestigial in the Renaissance, the modern period, and beyond. The heavens, for example, were not comprised of material clumps hurtling through space; they were spiritual entities that had direct correlates with parts of the human body. Physicians and astrologers were one and the same. If your zodiac sign was Aries and you were unwell, it meant that you should probably be bled from somewhere on your head; if you were a Cancer, then you should be bled from the chest; a Pisces, from the feet; and so on. The four elements, earth, air, fire, and water, bore relation to the four bodily humors, bile, blood, phlegm, and choler, which in turn formed a connection with the four corners of the earth and the four seasons. How these correlations were originally configured is entirely mysterious, but it is clear that these ordering systems and categorical frameworks ruled the day.

At the end of the sixteenth century in Perugia, Italy, a scholar named Cesare Ripa wrote a symbol-decoding manual called the *Iconologia* (fig. 3.3). This popular treatise (which over a period of more than two hundred years was published in multiple editions in Italian, French, Dutch, German, and English) provides us with a window into the symbolist mind. Here Ripa sets out the traditional correlations between visual images and their hidden meanings. He states that "[t]he figures that are made to express a thing different from that which we behold with our eyes, have no surer nor more common rule than the imitation of thoughts." And Ripa explains to his readers which visual images conjure which ideas and why.

The image of the lion, to use one of Ripa's more obvious examples, conjures magnanimity. A column or pillar is a code for emotional strength. And one can even symbolize the "Art of Eloquence" by drawing (or viewing) a sword and shield, "[f]or as these instruments defend the soldier's life and hurt his enemy; so the Orator, by his proofs, maintains his good cause and puts back the contrary party." Animals, in particular, were displayed and contemplated not for their own sake, but for some higher meaning. Ripa blended together ancient sources such as Pliny's *Natural History* with bogus hieroglyphic speculations in order to establish animal symbology. For example, the crane symbolizes vigilance because "the crane is supposed to sleep with a stone in its claws to use as a weapon if surprised." Or the goat, being "the most potent and easily aroused of animals," is a visual representation of "lewdness" (*impudicitia*). Or the stork is a manifestation of gratitude because (according to Pliny)

Figure 3.3. A symbolic image of "Lewdness" from Cesare Ripa's *Iconologia*. An "unchaste" woman reclines, fondling a blindfolded Cupid. A goat, which was considered the most potent and easily aroused animal, stands near the woman's foot. (TAKEN FROM THE HERTEL EDITION. BAROQUE AND ROCOCO PICTORIAL IMAGERY, TRANS. EDWARD MASER. DOVER PUBLICATIONS, 1971)

"it is very kind to its parents in their old age, building them a home, tending their plumage, and finding food for them."

The *Iconologia* is frequently mistaken as a narrow facet of art history, but it's really more of a cultural dictionary for understanding the dialogue between postmedieval Europe and the ancient Mediterranean. The rebirth of ancient culture during the 1500s meant the appropriation of that earlier symbolism, but this nascent symbol system inevitably subsumed medieval Christian symbology as well. It can be rather complicated in some cases. For example, an ancient pagan sculpture or painting of a shepherd carrying a sheep to market was a common decorative piece, but in the medieval period it became a Christ reference, and in the Renaissance it harked back to humanism again. Some things were fairly straightforward; portraying a lamb, for example, is a code for Jesus and his sacrifice. Bread and wine are not just bread and wine, they are the body and blood of Jesus. Water itself is a code for baptism and purification. Doves represent the Holy Spirit; peacocks represent paradise. A fish is doubly symbolic because Jesus made his disciples into fishers of men, but the Greek word *ichthys* also stood as an abbreviation for "Jesus Christ Son of God, Savior." The fish symbol has recently resurfaced in America on the car bumper of many a pious Christian. It has even elicited countericons of fish with feet that connote a Darwinian worldview. All of this meaning play is lost on the uninitiated, the person unfamiliar with the subtleties of how these symbols have taken on meaning in our culture.

So the taxonomy of nature as it appeared in imagery during this period was quite elaborate. The ordering of things into physical collections, however, was rarer, and it was not done for the purpose of piquing the viewer's curiosity (that would come later). The organizing of objects and symbols was undertaken for the sake of constructing some religious or otherwise supernatural narrative. If an object was actually displayed for public consumption, you could be pretty certain that it was a holy relic of some kind: a saint's finger or a prophet's tooth. There was so much of the "true cross" floating around in the early 1500s, Erasmus quipped that "if [all] the fragments were joined together they'd seem a full load for a freighter."

In the 1500s some budding naturalist-physicians collected specimens, but even these were usually explained and ordered by spiritual means. Aldrovandi, for example, engages in painfully prolix combinations of natural history and biblical analysis when he explains the mystical meaning of the rooster. The rooster is a symbol of the evangelical preacher, who brings not only the good news, but also the light of morning. He spends thirty-odd mind-numbing pages explaining how his understanding of the rooster's crowing behavior sheds light on the famous episode in the Gospels where Peter denies Jesus three time before the cock crows.

Ambroise Paré, a similarly spiritual naturalist, collected and studied birth defects, eventually writing a text, *Of Monsters and Prodigies*. The correlation of natural signs and supernatural meanings begins, in thinkers such as Paré, to become more causal, but his approach is symptomatic in that he categorizes the phenomenon of "conjoined twins" in three nonmaterialist ways. Conjoined twins can be exhaustively explained as the result of either (1) God's anger, (2) the devil's influence, or (3) maternal impression (when the fetus is affected by the mother's witnessing of a frightening person or event). So not only was the animal kingdom a cryptogram for spiritual reality, but the human form itself was a system of coded symbols.

The dioramas of Fredrick Ruysch (1638–1731) are well-known morality displays that communicate, in gruesome detail, the transitory and fragile nature of earthly life. In terms of historical chronology, Ruysch was actually exhibiting during the scientific revolution, but his dioramas are definitely in the spirit of an earlier age. Ruysch used organic objects, usually body parts that he meticulously dried and sculpted, to compose depressing scenes for his Calvinist patrons. Gallstones, kidney stones, and chunks of dried organs were the "landscapes" for infant skeletons posed in nightmarish scenes. The flesh is inherently morbid and fraught with suffering—atone for your sins.

Likewise, the Leiden anatomist Pieter Paaw (1564–1617) developed a moralizing display of morbid specimens. Paaw placed six human skeletons around the anatomy theater, each holding a pennant inscribed with phrases such as *Pulvis et umbra sumus* (we are dust and shadow). Historian William Schupbach describes the rest of Paaw's collection as containing "skeletons of a cow, cat, rat, ram and swan, and of an eagle with gilt talons. The pioneers of death, Adam and Eve, were represented by skeletons of a man and woman beside a tree."

Out of this phase of spiritualized classification, in which organizing and displaying nature explicitly aimed beyond nature itself toward the supernatural realm of significance, emerged the trend of encyclopedic natural history. *Encyclopedic* is a good term for describing the natural history that reigned from the mid-sixteenth to the mid-seventeenth century, just before the Royal Society began. Many of these naturalists, fellows such as Aldrovandi (1522-1605), Konrad von Gesner (1516–1565), John Gerard (1545–1612), Gaspar Schott

(1608–1666), and Francis Willughby (1636–1672), were "hinge" characters between the old spiritualist paradigm and the new empirical one. But they were all insatiably hungry for minute observations of nature, and as quickly as they took in information they recited back the loosely related observations. The mood during this phase seemed to be that if we could just observe the nuances of nature and re-create linguistic models of them, we'd have a science of nature. Being scientific at this time meant nothing more than assigning a name to your newly discovered plant or animal and then exhaustively listing descriptions under that name. After the dizzying double-edged (even triple-edged) meanings that were attached to nature during the spiritualist phase, it is small wonder that this new phase tried to concentrate on straightforward observation of the object.

The objective of these naturalists was thoroughness, and their guiding lights were structural morphology and, occasionally, animal behavior. But they gathered their data without any clear sense of a unifying system, and their zeal to be encyclopedic often resulted in the inclusion of bogus descriptions based on hearsay and mythical legends about the animals and plants. This combination of valid and bogus animal description was a time-worn tradition (very prevalent in the wake of the New World's discovery), and the seventeenth century was struggling to break from it. Gaspar Schott, for example, wrote a text entitled *Physica curiosa* in which he included illustrations and descriptions of exotic animals such as the camelopardalis (the alleged offspring of a panther and a camel). In his zeal for completeness Schott also included demons, monsters, freaks, angels, meteors, and more.

Aldrovandi's book on chickens includes a chapter entitled "Concerning Freak Chickens," wherein he sets forth credible embryological aberrations (e.g., Siamese chickens) together with incredible fantasy creatures (figs. 3.4, 3.5). At one point he states, "If any bird should be called a freak, it must be the rooster which I observed a few years ago, when it was alive, in the palace of the Most Serene Grand Duke Francesco Medici of Tuscany; it struck fear into brave men with its terrifying aspect." This conjures a wonderful image of burly grown men trembling in the presence of the dread chicken of Tuscany. These men of mettle, turned coward, beheld a creature with strange spikes protruding from its face—a creature with feathers "of such a form that they resembled scales over the entire body. Near the rump, where the tail grows forth, there was a round whitish tubercle. The tail was not made of feathers like that of birds, but was fleshy like the tail of quadrupeds—free of hairs, but with a lock at the extremity as usually is found in the tails of four-footed beasts."

Figure 3.4. A "freak" chicken from Aldrovandi's ornithology text. Two bodies united by a single head.
(ALDROVANDI ON CHICKENS, TRANS. L.R. LIND, 1963. BY PERMISSION OF UNIVERSITY OF OKLAHOMA PRESS)

Figure 3.5. A "freak" rooster with a quadruped's tail, from Aldrovandi's ornithology text.
(ALDROVANDI ON CHICKENS, TRANS. L. R. LIND, 1963. BY PERMISSION OF UNIVERSITY OF OKLAHOMA PRESS)

It was this sort of dubious freak-show material that particularly bothered the later Enlightenment naturalists. Buffon, for example, states that "Aldrovandi, the most laborious and most learned of all the naturalists, left, after a labor of sixty years, some immense volumes on natural history. . . . One could reduce them to one-tenth of their bulk if one removed from them all their useless parts and matters foreign to his subject." Buffon complains that Aldrovandi's natural history "is often mingled with the fabulous, and the author reveals in it too much of a tendency toward credulity."

Another case of rather gullible natural history can be found in John Gerard's *Historie of Plants* (1597). Here Gerard tells us of a remarkable plant, variously named "Goose Tree," "Barnakle Tree," or "the tree bearing Geese." At the time, a legend existed claiming that certain barnacles gave rise to adult geese, and Gerard not only reiterated the legend but also confirmed it wholeheartedly. He claimed to have seen with his own eyes a land plant release barnacles into the water, where they eventually grew to become geese! In sixteenth-century natural history texts this kind of fantastic entry could be found directly adjacent to very accurate and sober ones.

I've described these naturalists as "gullible," but that isn't entirely fair. We have to imagine a time when there were no clear guidelines for assessing the veracity of a naturalist's claim. Gerard, in the example above, seems to have been intentionally misrepresentational, but most sixteenth- and seventeenth-century naturalists had to take tremendous amounts of information from faraway places on faith—and they presented it to readers in good faith. The readers were not so much at the mercy of the naturalists as they were colleagues in credulity. Cross-referencing one text with another was tantamount to scientific verification, but, of course, if there was disagreement between texts, then the author with greatest prestige would win out.

The development of etching and printing technology gave the "descriptive sciences," such as anatomy, botany, and taxonomy, an amazing leap forward. Wood engraving began to proliferate in Europe around 1500, though the Chinese possessed this skill as early as the ninth century. Andreas Vesalius quickly set the standard for scientific woodcuts by publishing his collection of anatomy dissections, *De humani corporis fabrica*, in 1543. The natural historians quickly began to adopt printed illustrations into their catalogs, around the time that wood engraving gave way to steel and copper, which provided more potential for graphic detail.

The new illustration practices set up an interesting tension during the days of gullible natural history. On one hand, well-drawn images of an exotic animal such as a giraffe could provide standards (if the drawing was done from life) that might successfully constrain the more improvisational imaginations of the period. Detailed realism was a visual key for classifying organisms, especially since visual appearance was the reigning criterion of taxonomy. On the other hand, the cost of publishing new images and the nonexistence of copyright law led many naturalists simply to reproduce old etchings in the new texts or have old etchings serve as models for new drawings. If the original engraving was incorrect, then generations of copies would quietly foster the mistakes. In Gesner's *Historia animalium* he includes two drawings of the salamander. One of these is very realistic in terms of its rendering (proportions, markings, posture, and detailed morphology are all accurate); the other salamander looks like an embellished monster from a Hieronymus Bosch painting. The two images speak loudly and clearly of the differences between drawing from careful direct observation and drawing from hearsay. The latter was becoming less and less acceptable (figs. 3.6, 3.7, 3.8).

The case of Albrecht Dürer's rhinoceros drawing is an excellent example of the "standardization problem" facing natural history. Until Dürer's 1515 drawing of a rhino, Europeans were almost entirely unacquainted with the creature. Dürer's drawing is remarkable for its technique and beauty, and it effectively conveys the animal's massive weight and strength (fig. 3.9). Throughout Europe, Dürer's drawing was copied in natural history texts all the way up until the nineteenth century. The fascinating thing is that Dürer never actually saw a rhino in his lifetime. The famous Dürer rhino is based on a sketch and description of the animal that had been sent to Pope Leo X as a zoological gift. The ship carrying the animal capsized, and the poor beast drowned.

Even in the mid-eighteenth century, it was difficult for European naturalists to see the real thing. In 1749 Buffon traveled to the Saint-Germain fair to see a live rhino. He had an artist draw the animal and, using the new drawing, judged Dürer's famous drawing (reproduced everywhere) to be erroneous on several points. This illustrates the standardization problem nicely, because Dürer's image set a standard that reined in the fantastic flights of some naturalists. But Dürer's drawing includes a second horn at the shoulder of the animal that simply does not exist in normal individuals, and this nonexistent horn is reproduced like a bad gene in countless copies of the drawing. The second horn is a dead giveaway that a drawing has been copied from Dürer (fig. 3.10). Still, the addition of visual representation was a powerful contribution to the mania for description that characterized this phase of natural history. And it should be noted

Figure 3.6 (ABOVE, LEFT). A goat from Konrad Gesner's sixteenth-century *Historia Animalium*. (DOVER PICTORIAL ARCHIVE SERIES)

Figure 3.7 (ABOVE, RIGHT). A "Satyr" (believed to live east of India), from Konrad Gesner's sixteenth-century *Historia Animalium*. (DOVER PICTORIAL ARCHIVE SERIES)

Figure 3.8 (ABOVE). A "human-faced" carp from Konrad Gesner's sixteenth-century *Historia Animalium*. (DOVER PICTORIAL ARCHIVE SERIES)

Figure 3.9 (BELOW). Albrecht Dürer's "The Rhinoceros" woodcut, 1515. Notice the fabricated second horn at the shoulder, and the highly stylized armor-plate skin. (BRITISH MUSEUM)

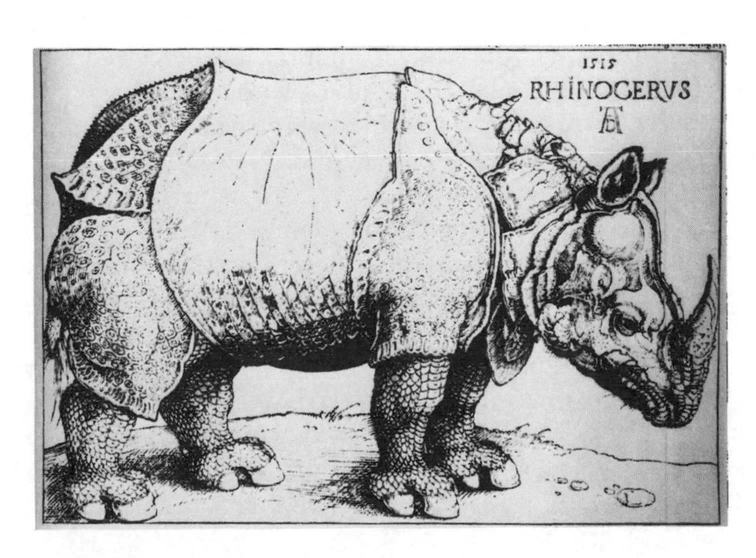

Figure 3.10. A rhino plate by J. Johnston. The image is clearly derivative of Dürer's famous work. Johnston's version appeared in a 1657 version of Gesner's *Historia Animalium*. (FROM IMAGES OF SCIENCE, BY BRIAN J. FORD, © 1992 BY BRIAN J. FORD. USED BY PERMISSION OF THE BRITISH LIBRARY)

that the proliferation of illustrated texts enabled, for the first time, writers and readers to treat these works as nature cabinets in book form. Obviously, collecting and arranging physical specimens continued and served as the basis for book contents, but it was now possible to spread miniature graphic representations of the collections all over Europe.

Something quite significant, however, was left out of these Byzantine descriptions of flora and fauna, namely, anything physiological. Besides the distastefulness of vivisection, an interesting feature of seventeenth-century science stood as a barrier to physiological exploration by natural historians. Physiology entailed the careful analysis of motion, but the study of motion was perceived to be the subject matter of physics. The forays into physiology of René Descartes (1596–1650), for example, were speculative and mathematical reflections on idealized machines. Natural historians of the time simply did not speak this language. The naturalists preferred to leave the issues of physiological mechanics to those savants who engaged in physics experimentation. Later both Cuvier and Hunter would eventually succeed in wedding these two estranged fields of study.

As the science of natural history developed, a tension began to grow between the impulse to exhaustively record every detail of an animal's or plant's structure, color, and habit and the impulse to bring an orderly and usable classification. The practice of analysis and description has a way of gorging itself once it gets started, and soon descriptions are stuffed with minutiae, none of which is useful. The system (if it can be called that) by which these naturalists organized their flora and fauna was frequently just alphabetical. While an alphabetical organization was better than nothing at all, it still did not express any meaningful relationships between the species themselves. Kinds of animals were categorized next to each other simply because their names started with the same letter. It was a fairly arbitrary form of order, and of course it would change drastically depending on which language was being used.

While this generation of naturalists was more directly interested in nature per se than their spiritualist predecessors, it would be a mistake to see the early 1600s as hosting the rise of a secular natural history. Naturalists such as Gesner, Aldrovandi, and Schott were all working within a theistic paradigm. Even the following generation of naturalists oriented their research toward the final end of God's glorification. Nehemia Grew (1641–1712), for example, writes a book with the economical title *Cosmologia Sacra: Or a discourse of the universe as it is the creature and kingdom of God. Chiefly written to demonstrate the truth and excellency of the Bible; which contains the laws of his kingdom in this lower world. . . .* And while this generation is even more attentive to material nature than their symbolist predecessors, people such as John Ray (1627–1705) still composed tracts such as *The Wisdom of God Manifested in the Works of Nature*. A truly secularized natural history would have to wait until the nineteenth century.

John Ray was an important player in this next phase of the taxonomy story. As we've already seen in chapter 2, the rise of the Royal Society in England and the Royal Academy in France signaled a growing institutionalization of the sciences in the late seventeenth and early eighteenth centuries. These institutions, comprised of thinkers such as Boyle, Hooke, and others, were consecrated to the new mechanical philosophy—the philosophy of microstructure. Nature, at the micro and macro levels, was understood as a machine. This fact, we will find, had important implications for the development of taxonomy.

As natural philosophy became more institutionalized, practitioners of a given science from all around Europe could talk to each other more easily, or at very least read each other's work more readily. This spirit of cooperation and communication led to many advances, but it also forced natural historians to face embarrassing problems. As naturalists became more familiar with their colleagues' work they recognized serious differences in their classificatory tables. John Tradescant, for example, grouped whales and even turtles in the "fishes" category. Following Aldrovandi, Worm, Cospi, and Grew all classed whales with mammals. René-Antoine Réaumur argued that slugs, snakes, and even crocodiles should be classed as "insects." As we have already seen back in chapter 1, John Ray argued that daffodils and tulips belong together in a group called "tuberous herbs," and Rivinus split the two into completely different groups. And no one really knew what to do with things such as sponges and fungi, since these seemed partly like plants but partly like animals.

In addition to these disagreements, naturalists (particularly botanists) had the continually nagging problem of balancing the different uses of their systems. The morphology of plants naturally seems to group them into categories of similar shapes and sizes, but arranging them according to their medicinal properties leads to a very different table.

These and many other such incompatibilities led the naturalists of this period to feel an urgent need for unification, and it expressed itself in the pursuit for a natural rather than artificial classification system. A very simple but intuitive way to grasp this distinction of natural versus artificial classification is to reflect on these two classes of things: (1) all dogs, (2) all things smaller than a breadbasket. These are both entirely coherent categories of things, but in the first case there is some essential relationship between the things grouped, whereas there is only an accidental relationship between the second group's members—a whiff of the artificial.

If the holy grail of systems could be found (the one that reflected essential connections with perfect resolution), then all differences of categorization would dissolve. Finding this totalizing abstract "cabinet," in which all individual organisms are perfectly categorized, and finding the methodological key to unlock the cabinet became natural history's full-time agenda during this phase.

An analogy may help to convey Europe's excitement for a natural system. If we think about visual arts in the late medieval and early Renaissance periods, we find the well-known struggle for optical perspective. Here we find the flattened but charming figures of Giotto, people painted bigger than buildings, halos that blot out figures' faces, twisted angles that would make Escher scratch his head, and characters comfortably violating the laws of space and time by appearing several times in the same scene. When Europe rediscovered Euclid, via the rich Arab tradition of perspective theory, Western visual art acquired foreshortening almost overnight.

It was the discovery of Euclidian optical theory, wherein space itself could be geometrized into a neutral and reliable grid, that allowed artists to place figures and structures relative to each other in an objective way. Soon after, one finds the visual depth of Brunelleschi, Leonardo, and Dürer, the last of which even built a little suitcase grid-tool for perspective accuracy while out in the field.

To be quite precise, Euclidian perspective is the natural system for organizing realistic visual art, not because reality comes packaged according to Euclidian rules (in fact, thanks to Einstein, we know it does not), but because in this case it is the natural system: Human eyes work in sympathy with Euclid's rules. Euclid may have thought that he discovered the way reality worked, but we know that he instead discovered the way the eye worked. In chapter 5 we will try to determine on which side of the subjective/objective line contemporary taxonomy resides. But discovering the natural taxonomic system in the 1700s would have been like discovering an objective perspective grid for all of nature. It was noble, it was inspirational, it was a pipe dream.

There are some important cultural presuppositions at play behind the very distinction of natural and artificial systems. The first and most important assumption is that a benevolent God is in his heaven. Remember that the cosmos was understood, at this time, as a giant machine. Clocks were the favorite metaphor of the day. If clocks have their orderly structure because intelligent minds have designed and built them, then the clock itself is a kind of proof of a designing mind. By analogy, the orderly cosmos is evidence for a designing God.

In understanding this era, one must keep in mind the omnipresence of Isaac Newton's influence. Newton was, of course, himself a major player in the Royal Society (its president, in fact, from 1703 to 1727), and scientists working in every discipline, including taxonomy, were intoxicated with Newton's amazing reduction of all motion to three or four laws. Using the universal language of mathematics, Newton had placed all of astronomy and terrestrial mechanics on objective and reliable foundations.

Taxonomists at the time, searching for their equivalent to Newton's certainty, were confident about eventually finding it, because while humans may be confused about the laws of classification, God surely isn't. But how can we be sure, ventures the skeptic, that nature isn't too complex for our human minds to grasp? The unflappable response was to point to the precedent of Newton's successful project.

This isolates the other extremely important assumption behind the quest for a natural system. The human mind works in essentially the same way as God's mind. Rationality, it is assumed, is the same everywhere, even in the spiritual realm. If not, we will be forever cut off from understanding our experience, and our science will be an elaborate sham. If our human mind (small and puttering) is not the same kind of "processor" as God's mind (huge and frictionless), then our pursuit of his great design is doomed to failure. From the scientist's perspective, it would be a cruel world without this isomorphism of rationality—not unlike telling a color-blind person on a scavenger hunt that the solution to the game can be found by following the blue signs.

As a corollary to this philosophical assumption, the taxonomists of this phase believed that there must be some universal language for expressing this eventual convergence of our minds and God's divisions of nature. After all, the physicists had done so well with their universal language of mathematics; overcoming geographical and linguistic isolation between scientists, eliminating ambiguities of native tongues, rising above the prejudices of local customs, and generally getting everybody together on the same page. Not only did taxonomists disagree about where to categorize an animal, but in many cases they could not even be sure that they were talking about the same animal, since the names for an animal could be different in various languages. The great Buffon, whom we'll examine in detail later, discussed this confused state of nomenclature. The attempts to get straight on the buffalo, for example, were tortuous.

The French word *buffle* was in circulation, and this animal was apparently prevalent in Greece and Italy. But the ancients made no reference to it. Aristotle did describe something called a "bonasus," but nobody really knew what that was. Aristotle's description of the creature in the *History of Animals* seemed to mirror the European bison, but then the description takes a strange turn. He claims that the bonasus "defends itself against an assailant by kicking and projecting its excrement to a distance of eight yards; this device it can easily adopt over and over again, and the excrement is so pungent that the hair of hunting-dogs is burnt off by it." Clearly Aristotle, like so many other naturalists, adopted the descriptions and fabulous tales of travelers into his natural history.

An anonymous Greek compilation of ancient natural history (circa the second century C.E.), entitled the *Physiologus*, was translated and embellished during the medieval period. These new Christianized versions of the *Physiologus* were known as bestiaries, and they were illustrated collections of beasts, birds, reptiles, and fish. The bonasus creature seems to have made it into a collection known as the Workshop Bestiary. Here the animal is described, in a way that is suggestive of the buffalo, as having the body and mane of a horse but the head of a bull. The creature is referred to as a bonnacon, and its fictional defense system has been amplified well beyond Aristotle's already humorous account. The Workshop Bestiary explains that "however much his front end does not defend this monster, his belly end is amply sufficient. For when he turns to run away he emits . . . the contents of his large intestine, which covers three acres. And any tree that it reaches catches fire" (fig. 3.11). By the time of the modern era much of this sort of factual embellishment was tapering off, but the naming confusions wore on.

The name *buffalo* was not Greek, nor was it Latin, but Julius Caesar had mentioned something called an urus, and Pliny had recognized the bison. At some point the urus and bison were combined under the name bubalus. Buffon states that "the confusion only increased with time: to bonasus, bubalus, urus, bison were added catoblepas, thur, the bubalas of Belon, the Scottish bison. . . . All our Naturalists have created as many different species as they have found names."

To end this sort of long-standing confusion in natural history, John Ray and John Wilkins called for a natural classification and a universal language. These two intellectuals were colleagues of Boyle, Hooke, and the other founders of the Royal Society. Ray was a brilliant observer, and he sought to unify his observations into a scientific system that could be used by other naturalists. An emerging issue for Ray's generation was whether a system could be useful only if it was true (meaning the one and only accurate mirror of nature). Not entirely clear about whether to pursue the path of the true (natural) classification or the useful classification (or whether these paths

Figure 3.11. The legendary bonnacon, from the twelfth-century Workshop Bestiary. The beast is portrayed here venting its flaming bowel in defense. (MORGAN LIBRARY, NEW YORK, MS M.81)

would converge), Ray contributed to his friend Wilkins's book project *An Essay Towards a Real Character, and a Philosophical Language* (1668).

Character here generally means a distinctive or significant mark, a graphic marking that signifies some literal fact. The term plays an important role in both linguistics and biology. *Character* is a technical term in taxonomy, and it refers to an attribute of an animal or plant. A diagnostic character is a special type of character that distinguishes a particular taxon. It is an identifying feature that tells the naturalist how to classify the organism. By analogy, someone who wants to know how to classify my particular computer may examine it to find clues or characters that indicate its species. One significant marking on the computer is a little rainbow-colored apple. This visual marking on the surface of the computer is a kind of diagnostic character that allows the observer to identify the species as Macintosh rather than IBM or some other species. To extend this analogy of the computer, we have to recognize that the little decorative apple symbol is helpful but also potentially misleading, because it could be just a decal improperly placed. Isn't there some more fundamental attribute or quality that distinguishes one species (and its members) from others? In this spirit, one can move beyond the exterior of the machine to the interior operating system and try to isolate the distinguishing character or characters. Here the cybertaxonomist is after the essence of the computer, the features that make it what it is.

Zoologists and botanists such as John Ray were attempting to find the same sort of essential characters for each kind of organism. As in the computer example, not just any character will do. This is why Wilkins, in his title, refers to a "real character." Just as in the case of the computer taxonomist who bases his classification on decorative decals, so too the naturalist might build a taxonomy on some trivial and unessential property (e.g., grouping animals according to their color). The ultimate goal, however, is to sift out the essential character, which, in a manner of speaking, reflects the organism's particular recipe.

Wilkins, using data from Ray and other Royal Society fellows, was attempting to discover a linguistic system that would reflect the real characters in nature. He was trying to find terminology that was unambiguous and application rules that were transparently rational, to unify the emerging scientific picture of the world. He wanted to undo the damage that had been done at the Tower of Babel. The universal grammar would be the means by which we humans could explore and express the natural classification. Actually, Wilkins was primarily interested in articulating the universal language first, and the naturalists' categories (such as those of Ray) were adjusted to fit with the tabulated grammar structure.

There was some disagreement about how to generate this universal language. Should the language philosophers immerse themselves in many different native tongues and then emerge with an artificial hybrid language? Or should philosophers eschew actual languages and pursue pure axioms and theorems of syntax and semantics by the light of reason alone? Interestingly enough, this pursuit of a universal language continued into the twentieth century. Broadly speaking, the Vienna movement of logical positivism was an attempt to discover and apply a universal grammar of all knowledge (Boolean algebra) by dint of reason alone. The twentieth-century version of the hybrid approach might be seen in Dr. L. Zamenhof's artificial language, Esperanto. Esperanto (which means "he who hopes") is a hybrid language that uses a compressed vocabulary based on root words from European languages. Unfortunately, Zamenhof's (who died in 1917) high hopes were dashed by the world's steadfast refusal to adopt his language. The high-water mark for Esperanto was a thoroughly unwatchable B movie entitled *Incubus* (1965), in which all dialog is in Esperanto. *Incubus* stars a young William Shatner struggling with forces of good and evil—as well as with a completely awkward artificial language.

In John Ray's contribution to Wilkins's treatise on language, he tried to get the plant world to conform to the logic of Wilkins's grammar. But in the end Ray regretted the classification, because several plant groups were distinguished by accidental properties such as taste and smell. The onion and leek, together with the hyacinth and the iris, were categorized under the class Bulbosae. But the first pair were subdivided from the other pair, using the criteria "strong-smelling" versus "weak-scented." While this human-centered way of classing organisms may have played well enough in earlier ages, it did not seem very scientific by the new standards, and Ray knew it. Dividing nature according to its smell seemed like gerrymandering. The scientific standards of objectivity that were emerging at this time sought to remove the biased human subject from the empirical findings, but here was the human nostril at the very center of nature's plan.

Continuing his pursuit of better taxonomy, Ray broke with his earlier system and in his *Historia plantarum* (1686) argued that the defining character of a plant group was its sexual equipment. By analyzing the reproductive parts of plants, he was able to arrive at reasonable categories for kinds of plants. Some plants had flowers, some didn't; some plants had regular corollas, some didn't; some plants had naked seeds, some didn't; and so on. This reproduction-based system of organization was a more natural way (and a more scientific way) of grouping organisms. It seemed as if the holy grail of taxonomy might be realized, because classification might be achieved by the study of a single character (or a very finite list of characters). The character of reproductive organs could be the key to organizing the mysteries of nature.

At this promising moment in history, however, things began to look much more complicated and discouraging for the study of natural history. Remember that the bandwagon philosophy that was overtaking Europe (with good reason, given its successes) was mechanism. And at first this matter-based philosophy was completely consistent with the important assumptions that God designed the world and that the human mind could grasp or interpret the design. But unsettling implications began to arise out of the new mechanical orientation.

To understand natural substances better, to see what makes organisms and minerals tick, the founders of the Royal Society pursued the microstructure of things (fig. 3.12). Remember that in chapter 2 we encountered Robert Boyle and Robert Hooke writing treatises such as *Origins of Forms and Qualities According to the Corpuscular Philosophy* and *Micrographia*. In these early works of chemistry they showed how various crystal shapes were produced by different arrangements of spherical particles. This seemingly innocent move was in fact rev-

Figure 3.12. A detail of Robert Hooke's "flea" drawing, from his 1665 *Micrographia*. The drawing was made using an early compound microscope. (FROM IMAGES OF SCIENCE, BY BRIAN J. FORD, © 1992 BY BRIAN J. FORD. USED BY PERMISSION OF THE BRITISH LIBRARY)

olutionary in its implications. The revolution consisted in overthrowing an Aristotelian view of natural substances.

Everything you see in nature, Aristotle argued, is composed of two basic principles, form and matter. Look around. Dogs, ferns, oak trees, cockroaches, sharks, coffee cups, and human beings are all composite creatures. They are combinations of material stuff (flesh, wood, clay, water, whatever) and formal organization (the blueprint, so to speak). A cup and a saucer are made up of the same basic stuff (clay), but it is the form (blueprint) of the coffee cup that distinguishes it from the saucer (which has its own form). Only when the lump of clay is given the cup form (an open cylinder with a closure on the bottom) do we have a cup. Everything in the natural world, Aristotle argued, is composite in this way. If you want to scientifically understand a dog, or a fern, or a coffee cup, you must look beyond the thing's material stuff to its form. The form of the cup is its essence, and this is intuitively clear when we realize that we can make a cup out of a variety of materials (tin, ceramic, wood, etc.) if we know the form. The form or essence is what makes a thing what it is.

Just as the cup has its essential form, so too, Aristotle argued, plants and animals have their essential forms. And just as it is the essence "cup" that makes this hunk of clay different from that hunk (say, a saucer), it is the essence "human" that makes this hunk of flesh different from that hunk of flesh (say, a chimp). Following Aristotle, then, the natural historian was out to discover the essence of each animal. The pursuit for a natural classification was in large part a quest to sift out the essential character of an organism. Once you had discovered the essence of the dog or the buffalo or the whale, your science was on unimpeachable grounds. Your science would be eternal and changeless, because while matter may come and go, essences never change.

It was from this Aristotelian tradition, deeply embedded in the culture, that early taxonomists pursued the defining single trait (or character) of each organism. This led to the idea of weighted characters. If each animal consisted of essential properties and accidental properties (idiosyncrasies of matter), then the taxonomist should weigh the essential properties more heavily than the accidental ones. Weighing the reproductive organs more heavily than tooth structure, for example, meant that the taxonomist conceptualized reproductive equipment as more crucial than dentition to the very existence of the animal. The sexual system of an animal revealed more about the Deity's design (the original immaterial form of the particular animal).

There was a serious problem with this essence-seeking modus

operandi. No one had actually found an essence for an animal. There were suggested features, but nothing very compelling. Every time some naturalist defined an animal by its sexual organs, some other naturalist would point out that the means of locomotion was an equally essential feature, and so types of feet and wings and flippers should be weighted more heavily. Very quickly another naturalist could point out that the means of nutrition was really the essence of the animal, and so mouth parts, gullets, and so forth should be weighted more heavily. The following thought experiment may give you a sense of the difficulties posed by this weighting problem. Imagine that you are an alien from another planet, come to earth for the first time. As an alien, you do not already possess a prejudice about how earth animals should be grouped. You're given five creatures: a bat, a dog, a robin, a mosquito, and a penguin. And you are given four characters by which to classify them: beaks, wings, hair, and parturition (birthing process). Try the experiment using one character at a time. You discover that each character used carves groups differently. Some rather weird groupings occur, but then, this is the rub. To the alien you, none of the groupings could seem weirder than another; none of them could seem more accurate than another. Without some background theory or conviction about which animals should be grouped together, you are not sure which character to weigh more heavily. The most that the early taxonomists could hope for was that their divisions would not be totally incongruent with those of the folk herbalists and farmers (and, of course, those folk classifications were just based on human-centered needs and preferences).

What actually overthrew this Aristotelian thinking, however, was not any internal incoherence, but the eventual arrival of an explanatory alternative. Boyle and others could persuasively show that the thing that made one crystal different from another, or the thing that made an apple different from an orange, was not some mysterious essence. It was the arrangements of the microparticles that defined what sort of a thing it was. Through his distillations and reductions Boyle was actually transforming some substances (liquids, as we saw earlier) that were supposed to be essentially constant. Subsequently, everything from apples and oranges to chimps and humans no longer appeared to be homogeneous material stuff differentiated only by immaterial forms. It was an issue of selective emphasis, but for the Royal Society's founders, matter counted in a way that it had not previously. Compared to this new way of analyzing nature, the idea of an essential form seemed scientifically sterile. Perhaps, the mechanists pondered, there was no such thing.

All of this had a profound effect on John Ray. Pursuing a natural classification is tantamount to pursuing the essences of plants and animals. But if there are no essences, if the differences between organisms can be explained by how their microparticles happen to clump together, then trying to say that one character is more essential than another is completely misguided. This metaphysical tension leans away from the idea of a designing God that specially creates each kind of organism. The impious dimension of metaphysical atomism, however, was almost unthinkable to the founders of the Royal Society. And if the new chemistry had been the only worrisome thing, thinkers such as Ray would have successfully ignored it and gone on pursuing the essential character.

Another feature of the corpuscular philosophy, however, was too difficult to ignore, and it set John Ray on a path of skepticism that Buffon would eventually continue. One of the implications of the new physics and chemistry that were emerging is the relationship between the human mind and nature. In Aristotle's day, one assumed that by using one's senses, one could examine a plant or animal in detail and feel confident of learning the unmediated truth about the creature. But in the modern period (particularly after Descartes) natural philosophers were not as confident in their own senses. Our five senses frequently lie to us, or at the very least distort reality. We have to be extremely careful about assuming that our experience is giving us an undistorted picture of reality. To use a time-worn example; the water in a bucket feels hot to Mary's touch but cold to Anna's. But (as we've already seen) the water cannot be both hot and not-hot. Therefore, we conclude that the property "hot" exists *in Mary* (her sensory apparatus and brain) and the property "cold" exists *in Anna*. "Hot" and "cold" are qualities that arise in us when human sensors come in contact with certain molecular motions, but they are not qualities that inhere in the water itself. And while it is true that slow-moving molecules will cause an experience of cold in Anna's hand, it would be a confusion to think that slow-moving molecules *are* cold, since we can easily imagine other life forms feeling very comfortable in waters that we would find unendurable. No one who hates the taste of Gorgonzola, for example, thinks that badness or pain exists *in* the cheese. These are subjective experiences.

From this insight we know that any input into our senses is going to be altered by those senses. The intellectuals of this period made a distinction between primary and secondary qualities. The secondary qualities of the Gorgonzola are its color, taste, texture, and size. The primary qualities of the Gorgonzola are the motion, quantity, and geometrical arrangement of its microparticles. So primary qualities are invisible to us, and secondary qualities make up our daily experiences. According to Robert Boyle and (most famously) John Locke, these primary qualities cause (but are not the same as) our secondary qualities. This is more straightforward than it sounds. The invisible microstructure of garlic, for example, certainly causes the smell we perceive, but that odor is not itself a primary quality. This is the old tree-falls-in-a-forest thing. If you are cooking up some garlic and all perceiving life forms are suddenly wiped out in a nuclear holocaust, can there be such a thing as garlic odor? No. There will be specific garlic oils converting to steamy airborne particles, but there will not be odor. The lesson of all this is that human beings, like other perceivers (animals), get access to the world only in a mediated form; reality is always filtered for us by our own biological equipment. Human beings are attuned more to certain experiences and less to others, but what a scientist trying to understand the natural world on its own terms wants is to get access to those primary qualities and thereby avoid the biases of her own ideology, race, gender, species, and so forth.

The reason why all this became problematic for taxonomists such as John Ray is that they started to wonder whether they could ever get at a defining character. In observing the plant, Ray experienced many different traits; they had seemingly important colorations and smells, they had specific tastes and textures, and so on. Yet if Boyle and Locke were right, these traits were just subjective perceptions caused by unseen microparticles. Dispensing with these subjective traits as unscientific, he had settled on the reproductive organs as certain indicators for classification. But if you were measuring and counting and recording the specifics of a plant's sexual parts, then you were perceiving them. And we've already seen that perception is always a subjective experience. From a logical (and epistemological) point of view, then, a plant's smells and the number of pistils it has are equally subjective. How can we be sure, once we've settled on sexual parts, that we haven't chosen some totally irrelevant feature of the plant? Defining the organism by one character seemed reasonable when we felt confident that there were essences and we could directly perceive them. But now, in the grip of skepticism, it seemed arrogant and naive to think that a reproduction-based classification was the true mirror of nature. It *could* be correct, but if we could never get out from behind the veil of secondary qualities, how would we ever *know?*

From this struggle with philosophical skepticism, Ray emerged with another amended taxonomy procedure. If we could not see essences and we could not know which characters to weight, then we must abandon the search for one specific character. Instead, we would do better to record many distinguishing characters and attempt a holistic interpretation of the organism. Rather than looking only at one feature, we should look for overall similarities between organisms. In this way we run less risk of being duped by one overemphasized subjective perception. If they are all subjective perceptions anyway, then the odds have it that the more traits we take into account, the more likely we are to properly classify the specimen. Locke believed that the "frequent productions of monsters, in all the species of animals, and of changelings, and other strange issues of human birth," taken together with the fallibility of our own senses, signaled the impossibility of a "certain science" of classification. The boundaries of species and genus (where one kind of animal is demarcated from another) were, according to Locke, simply agreed-upon human conventions, nothing more. In this way of thinking, Chihuahuas and Great Danes are classed together because we say so, and cats and squirrels are classed apart because we say so. From God's perspective (presumably, the truth about reality), the Great Dane and the squirrel may comprise the true category.

Notice what happened here. The assumptions that undergird the search for a natural taxonomy were changed for Ray. He was still quite certain that God designed and created the species around us. But he no longer believed that our human minds (and our senses) can completely access this information.

Remember that this assumption of an isomorphism between the divine and the human mind was a driving force behind Enlightenment natural history (perhaps a driving force behind all science). The later naturalist Louis Agassiz, whom we encountered in the race/species debate, nicely expressed this earlier mind-set himself. He said that taxonomy systems were "translations into human language of the thoughts of the Creator. And if this is indeed so, do we not find in this adaptability of the human intellect to the facts of creation, by which we become instinctively . . . the translators of the thoughts of God, the most conclusive proof of our affinity with the Divine Mind?"

John Ray was not so sanguine. The best we can hope for, he thought, is a probability that our classification (made up of secondary qualities) matches the real divisions in nature (caused by inaccessible primary qualities). Ray believed that only the afterlife could give us the unmediated epiphany of the real natural classification. He states, "The correct and philosophical division of any genus into species is by essential differences. But the essences of things are unknown to us. Therefore, in place of these essential characters, some characteristic accidents should be used, which are present in only some of the species and in all the individuals contained in these." Instead of classifying plants according to their flower and fruit structures only, Ray points out that we might as well include root structures and stem structures. All the characters are weighted equally now.

One last point about Ray and similar French naturalists is in order here. Around 1700 a slight shift was occurring in naturalists' perception. Recall that the awe and wonder early curiosity cabinets engendered in their viewers was based primarily on an encounter with the dizzying diversity of nature (also implicit proof of God's fecundity). But now, as some system began to shape the armoires, the awe and wonder of the nature enthusiast was engendered by the striking adaptations in nature rather than its dumb diversity. Ray led the way by exposing his readers to the ingenuity of nature, its repeated correlation of structure to function. These were the beginnings of the theological "argument from design" (in its modern version, anyway) that we encountered back in chapter 2. The mole, for example, has been designed or adapted perfectly for life underground: Because it is blind and lacks true eyes, dirt will not interfere with vision; notice how its hands are like large shovels for digging. A bird needs to be of light weight for the purposes of flight, and lo and behold, its bones are hollow and lighter than those of flightless creatures. These and countless other details of adaptation, according to naturalists such as Ray, speak of the Creator's wisdom. It is a little ironic that while Ray became skeptical about our perception of an essential character, he did not extend his skepticism to the perception of adaptation in nature. Perhaps that ice was just too thin.

Following John Ray's lead, the French had their own homegrown natural theologians in the form of Abbé Pluche and René de Réaumur. Réaumur contributed to Pluche's extremely popular work *Le Spectacle de la nature*, a work that, like Ray's, sought to glorify God's ingenuity by displaying the awesome cleverness of nature. Efforts such as those of Pluche and Réaumur led to an increase in pious bourgeois curiosity cabinets. While the average person was still collecting for diversity, the scholars were starting to think more about adaptation and function. Pluche and Réaumur believed that the spectacle of nature was an entertaining and edifying experience but that the complex details of science were beyond the powers of the dim-witted public. Pluche says: "We believe that it is more suitable for us to describe the exterior decoration of this world and the effect of the machines that produce the show. We are allowed to see this much. We even see that it has been rendered so brilliant only in order to pique our curiosity. But, content with a representation that fills our senses and our intelligence, we need not ask that the engine room be opened to us." Two things are suggestive about this quote. One is the allusion to that skeptical impasse that Ray experienced—perhaps the engine room can never be opened to limited human minds and senses. Two, this is an early version of our edutainment issue. Is the best that can be hoped for in a science education a curious encounter with the surface spectacle of nature? Réaumur believed that the engine room of nature could be accessed by professionals but that the average citizen would find it all much too dry. Better to spare them the disappointment of trying to move past the spectacle. As we have already seen (and will continue to explore), this issue of edutainment is at the very heart of museology. Here Réaumur and Pluche are raising the issue as authors of popular science texts, but the same question is nascent in museum practice.

During the early part of the eighteenth century amateur and professional collecting were still united in their aesthetic response to nature, but increasingly the professional collector began to define himself by the new interest in secondary causality. Organizing items according to aesthetic taste was fine for the novice, but for those who were initiated into the subtleties of the new mechanical philosophy, the organization of specimens became preparatory work for further scientific inquiry.

CHAPTER 4

Taxonomic Intoxication, Part II: In Search of the Engine Room

LOOKING OUT THE DUSTY WINDOW of the Galerie d'anatomie comparée, one can see the giant botanical hothouse and the multicolored "carpeting" of blooming flowers and herbs. It was somewhere out there, in the sculpted Jardin des plantes, that a small flock of new students hovered around a professor named Bernard de Jussieu. The year was 1738, and the Garden of Plants was then called the Jardin du roi. Jussieu's official title at the garden was "subdemonstrator of the exterior of plants" (which was considerably more prestigious than it sounds), and he was taking a group of students through some botany drills. Jussieu must have smiled as he held up one plant in particular. Turning to the students, he asked them whether they could identify this utterly rare and unfamiliar specimen. He probably paused dramatically, knowing full well that no one would be able to identify the obscure plant. Being a college professor myself, I know all about this useful pedagogic technique. You ask some thoroughly impossible question, and then you wait for the crushing silence to humble all the students into studying harder. It's very effective, as long as you don't use it too often.

On this day in 1738, however, there was no silence after Jussieu's impossible question. From the back of the small crowd came a steady voice that provided the precise answer. The awed students turned to stare, and Jussieu gathered himself. Gazing astonished at the new face in the crowd, Jussieu said, "You, sir, are either the devil or Linnaeus." It was, in fact, the brilliant Carl von Linnaeus (1707–1778), who, at the time of this episode with Jussieu, had already published the *Bibliotheca botanica*, the *Genera plantarum*, and a short version of his eventual masterpiece *Systema naturae*.

It would be very difficult to overestimate the impact that Linnaeus had on all branches of biology. A Lutheran minister's son from Sweden, Linnaeus was utterly devoted to the project of a rational and usable taxonomy that would both advance the science of nature and glorify the Creator. In 1732 the University of Uppsala commissioned him to study the plant life of Lapland. He set out with his "light coat of linsey-woolsey, leather breeches, a round wig, a green leather cap and a pair of half-boots" carrying a microscope, telescope, light shotgun, and drawing equipment. Before it was over, Linnaeus had walked some four thousand miles, made accurate first-hand descriptions of important flora, and established an impressive collection. He named more than 9,000 plants, 820 shells, 2,000 insects, and 470 fish. It's hard to imagine the determination and patience required for this achievement when most of us can barely distinguish between a millipede and a centipede.

Generally speaking, Linnaeus was rather unfazed by the skepticism that tortured thinkers such as John Ray. Either Linnaeus did not appreciate the objections to the assumption of isomorphism between God's mind and the human mind, or his thoughts were something along the lines of *To hell with it—I'm going to pretend that taxonomy can read God's mind and just see how far I get.* Either way, with a kind of get-things-done attitude, Linnaeus pushed ahead with Aristotelian logic under his arm, beating back the tendrils of taxonomic confusion.

The taxonomic distinctions of species and genus were already in wide usage by the time that Linnaeus came onto the scene, but the Swede added the variety, which is one rank below the species level, and the order and class, which are two levels above the genus rank. This new five-level system turned out to be very helpful to naturalists. If a naturalist had to memorize every organism's local folk name, he would need superhuman powers of recollection. The nested categories helped with the practical problems of proper specimen identification. It is actually the practical advantages of Linnaeus's system, rather than the theoretical claims, that put so many biologists in his debt.

His crowning achievement was his invention of the binomial system for naming taxa. Remember the chaos of the earlier phase, with names multiplying everywhere. Linnaeus actually delivered a workable universal language. It was just the old universal language of Latin in new clothing, but the logic for applying the terms was transparently functional. The way to designate a kind of plant or animal was to conjoin the specific name to the generic name; thus the designation is binomial. *Homo*, for example, is the generic name, and *Homo sapiens* is the specific name. Varieties are like races, and Linnaeus lists the following varieties of mankind in his *Systema naturae:*

americanus, *europeaus*, *asiaticus*, and *africanus*. But interestingly enough, there are some other varieties listed under the *Homo* genus in the tenth edition: *Homo sapiens sylvestris*, also known as "forest man" or troglodyte (better known as the orangutan); *Homo sapiens ferus* (the wild child); and *Homo sapiens monstrosus* (humans with various distortions of the body). It's clear from this listing that *variety* had a looser meaning than *race* (though it's admittedly hard to imagine a looser term than *race*).

The wolf, according to Linnaeus's system, should not be discussed using vernacular terms such as *loup*, *lupo*, *wolf*, and *lobo*. To be scientific, the animal should be placed on a map of ascending categories: species, *lupus;* genus, *Canis;* order, Carnivora; class, Mammalia. Likewise, the honeybee could be placed on the map this way: species, *mellifera;* genus, *Apis;* order, Hymenoptera; class, Insecta. The greatest demonstration of the effectiveness of Linnaeus's binomial system is the fact that we are still using it. We have even preserved four of his five categories in our contemporary taxonomy, though we have added the categories of family, phylum, and kingdom for greater clarity.

Assigning a name was a quasi-religious practice for Linnaeus and his followers. Like a more sophisticated Adam, the first taxonomist, Linnaeus used language to harness nature's chaos. He explains the importance of the name in the introduction to his *Systema naturae:*

> Man, the last and best of created works, formed after the image of his Maker, endowed with a portion of intellectual divinity, the governor and subjugator of all other beings, is, by his wisdom alone, able to form just conclusions from such things as present themselves to his senses, which can only consist of bodies merely natural. Hence the first step of wisdom is to know these bodies; and to be able, by those marks imprinted on them by nature, to distinguish them from each other, and to affix to every object its proper name. These are the elements of all sciences; this is the great alphabet of nature: for if the name be lost, the knowledge of the object is lost also; and without these, the student will seek in vain for the means to investigate the hidden treasures of nature.

In addition to the binomial naming system, Linnaeus offered up rules for identifying and sorting organisms. Actually, these were not new rules. Again, either Linnaeus had not heard that Aristotelianism was in trouble, or he had heard it and didn't believe it. He took Aristotle's old method of logical division and pressed it into productive service. According to the logic, there were five important logical dimensions of each organism: the definition of the organism, the differentia, the genus, the properties, and the accidents.

To understand this system, let us apply it to the case of Steve Asma. The definition of Steve will be the formula that expresses the essence of his species, *Homo sapiens* (for Aristotle, one cannot define an individual, only the species of which that individual is a member, because definitions are comprised of changeless propositions, whereas individuals are fluctuating material entities). What is the essence of human being? The answer is right there in the name; *sapiens* = wisdom, discernment, rationality. So the definition of Steve Asma is "rational animal." When Linnaeus employed the term *genus*, it designated just one step up of generality from the species. So the genus of Steve is "human being," *Homo*. The differentia are the traits or diagnostic characters that make up the definition and therefore set aside one species as different from another species within the same genus. They are the building blocks of the definition, and in this case the differentia would include "rationality." In some places (and we noted this back in chapter 1), Aristotle offers other characters for humans, such as "featherless" and "bipedal," which if taken together establish more complete differentia for categorizing. Now, the properties of an organism are the features that flow directly from the essence. A property of Steve is "ability to use language" (even talk about himself in the third person), because this is a consequence of his being a rational animal. An accident of Steve Asma is a quality or feature that has no important connection to who and what Steve is. "Parts hair on the right side" or "is hungry right now" are features that are accidental to Steve Asma. He would still be Steve Asma without these things.

Linnaeus takes this Aristotelian logic and transforms it into a workable research program. The first thing he does with regard to plants is commit to the single-character method of identification. Like John Ray's focus on the sexual organs, and unlike Ray's later holistic approach, Linnaeus believed that the flowers and fruits of plants would lead the way. All plants, he reasoned, are in the process of preserving themselves as individuals, but also preserving their own species. The vital activity by which they preserve themselves is the activity of nutrition, and the activity by which they preserve the species is reproduction. So it makes a certain kind of sense that if you want to make a genus designation, you should examine the reproduction equipment—the structures of the organs.

All told, Linnaeus divided plants into thirty-one sexual characteristics. Fruit consists of the preicarp, the seed, and the receptacle, while flowers are dissected into calyx, pistil, corolla, and stamen. Each of these structures can be further subdivided. Stamens and pistils each submit to three divisions, seed and receptacle each reveal four more parts, the preicarp has eight internal units, the calyx has seven, and the corolla is comprised of two parts. As if this wasn't enough chopping, Linnaeus showed how each of these thirty-one parts could be analyzed using the criteria of shape, position, proportion, and number. It seems unwieldy, but the system worked amazingly well. By analyzing the number and position of stamens, the naturalist designated the class of the organism. By assessing the number of pistils, the plant could be assigned to an order. The genus was determined by a comparative study of all thirty-one parts together, and the species designation incorporated some of the structures of the nutrition function. Linnaeus had constructed a very good taxonomic toolbox.

The tremendous observational data that he amassed, together with the logical system of anatomy and the binomial nomenclature, led many naturalists to proclaim that they had finally found their Newton. To the extent that the Linnean system was useful for organizing and identifying organisms and unifying the scientific language of natural history, Linnaeus was a Prometheus of sorts. By developing workable categories and a calculus of sex organs, he gave natural history a foundational framework for the deductive identification of organisms and the inductive development of kinds. And he was even closer to Newton in that he did not pretend to know or speculate about the causes of nature's taxonomic arrangement. Certainly God was always the distant cause, but the proximate cause was simply not addressed. Linnaeus was the great systematizer of surfaces only. He organized the spectacle of nature better than any predecessor had, but he was not interested in the engine room.

It goes without saying that the Swede was an avid collector. From his childhood days (tending his father's garden) to his final years (when students continued to send exotic specimens from their various expeditions), Linnaeus worked on his elaborate collection. In addition to Linnaeus's influential methodology, his collection of specimens thoroughly advanced the science of natural history. It was this cabinet that led directly to the founding of the Linnean Society in London in 1788. The Linnean Society is perhaps most renowned for hosting the two famous papers in July 1858 of Charles Darwin and Alfred Russell Wallace on the discovery of natural selection.

Shortly after Linnaeus's death, in Sweden, a young student named James Edward Smith was breakfasting in London with Sir Joseph Banks, then president of the Royal Society. Banks, who was a friend and supporter of John Hunter, had made his name with Captain Cook in the 1768 *Endeavor* expedition (to witness the transit of Venus across the sun). During this breakfast, Banks informed Smith that he had just been offered the entire Linnaeus collection (all manuscripts, books, correspondence, and specimen cabinets) for 1,000 guineas. Unable to fund the dowries of their four daughters, Linnaeus's widow was compelled to sell off the collection. To Smith's astonishment, Banks was declining the offer. But James Edward Smith no doubt envisioned his own career rocketing upward at the prospect of possessing the Linnaeus collection. So, with no money of his own, Smith harassed his reluctant father into coughing up the capital. In 1784, six years after Linnaeus's death, a ship pulled into England carrying the collection in twenty-six cases. The ship had narrowly missed a high-seas interception along the way, for when the transaction between Smith and Linnaeus's widow had taken place, the king of Sweden, Gustavus III, was away in France, and when he returned to find that the collection had just left his shores on its way to England, he was outraged and sent a ship to intercept the collection. The Swedish vessel never caught up, apparently, because the specimens were in Smith's hands that year, and he was elected a fellow of the Royal Society the following year.

Smith was by no means just a greedy collector or social climber. He quickly used Linnaeus's collection as the founding database for the new Linnean Society, devoted exclusively to natural history. In his introductory comments in the first *Transactions* of the society, Smith said, "I consider myself a trustee of the public. I hold these treasures only for the purpose of making them useful to the world and natural history in general." Currently the London-based collection is off-limits to the public. The society is trying to conserve the material by housing it in atmospherically controlled chambers. Professional botanists and biologists are still allowed to work with the collection, and one can only imagine Linnaeus's complete approval of his cabinet's longevity. In addition, an international team is currently editing and organizing Linnaeus's seven thousand letters for publication on the World Wide Web. The Internet publication will include images of rare plant specimens from Linnaeus's collection, and public access to this cybercabinet promises to be unprecedented.

An important historical feature of the Linnaeus collection and subsequent Linnean Society is that it represents a development of autonomy for the science of natural history. Using the collection, Smith broke away from the Royal Society and argued that natural history now required its own institution. The Royal Society had developed under the posthumously realized guiding vision of Bacon, whose "Solomon's House" ideal had placed all forms of knowledge and inquiry together under one roof. It was not uncommon during the Enlightenment for natural philosophers to work on astronomy issues one week and chemistry issues the next. But the nineteenth century began to see increased fracturing of the disciplines, with more and more specialization. Smith explains in the *Transactions* why the new, specialized Linnean Society would be beneficial: "It is altogether incompatible with the plan of the Royal Society, engaged as it is in all the branches of philosophy, to enter into the minutiae of natural history; such an institution, therefore, as ours is absolutely necessary, to prevent all the pains and expense of collectors, all the experience of cultivators, and all the remarks of real observers, from being lost to the world." This general trend toward autonomy of disciplines was received with some trepidation by important members of the Royal Society, and while the breakaway worked smoothly for the natural historians, it led the president of the Royal Society, Joseph Banks, to oppose the breakaway of an Astronomy Society on the grounds that it would severely damage its mother institution. Ultimately there has been no stemming of the tide of specialization in the sciences, and our own age is marked both by advances through specialization and by a notable inability of scientists to talk to each other and the general public.

The last two figures that must be examined briefly before we can understand the respective collections of Hunter and Cuvier are two iconoclastic Frenchmen named Georges-Louis Leclerc, comte de Buffon (1707–1788), and Jean-Baptiste Lamarck (1744–1829). Buffon and Lamarck both developed revolutionary ideas about taxonomy, ideas that eventually challenged Cuvier's museum display tactics. Hunter's museology independently parallels in some ways many of Buffon's more explicit theories of nature. Sketching those theories will help us better understand Hunter's displays as well as Cuvier's alternative museology.

Buffon was a rebel at heart but managed to negotiate the political and institutional nuances well enough to become the director of the Jardin du roi. His collected works, totaling thirty-six volumes, were hugely popular throughout Europe, and he was more widely read than even Rousseau and Voltaire.

Buffon saw Linnaeus's classification as an elaborate human invention that, while reasonably effective in parochial domains, was ultimately false. Those who followed the Linnean system, Buffon believed, were seduced by the clean logic of it, but they had become blinded by their own system and could not see the true subtleties of nature.

Buffon pointed out many places in Linnaeus's taxonomy where groupings violated our commonsense intuitions and the age-old intuitions of herbalists and animal breeders. In Linnaeus's system, for example, elms and carrots were classed together, as were roses and strawberries. Buffon claimed that Linnaeus's obsession with classification had also led to the absurd zoological grouping of man, monkeys, sloths, and scaly lizards under the same category, Anthropomorpha. Also, hedgehogs, moles, and bats were grouped together as ferocious beasts (Ferae), and hippos and shrews were classed together as beasts of burden (Jumenta). As Linnaeus moved from botanical to zoological territory, the clean divisions grew messy, and he was forced to consider more than one character for classification. Choosing characters became somewhat arbitrary, but Linneans, still confident from the successes in botany, refused to see their own choices as arbitrary. Buffon was there to point out their errors.

In the "Premier Discours" of his *Histoire naturelle*, he famously criticized the attempt to find discrete categories in nature: "Nature moves through unknown gradations and consequently she cannot be a party to these divisions, because she passes from one species to another species, and often from one genus to another genus, by imperceptible nuances." This provocative statement is not about species shading into one another *over time;* it is not an evolutionary statement. Buffon is claiming that the categories of organisms living now are not mutually exclusive. Remember that the Linnean/Aristotelian method sought to establish a defining character that all the members of one species possessed but no member of another species did. In this way one could apply the either/or logic to species—either you had rationality or you didn't, either you had webbed feet or you didn't, either you had canine teeth or you didn't. While this simple use of the either/or logic is trivially true, the messiness arises when you try to stipulate a short-list combination of traits to classify real organisms. This list of traits must designate members and exclude nonmembers. Buffon pointed out that these defining short lists, while helpful at first, were frequently violated by the existence of intermediate animals.

The shrew, according to Buffon, was an intermediary between the rat and the mole, sharing traits with each species (fig. 4.1). The pig also seemed to fall somewhere between the whole-hoofed and cloven-hoofed animals (fig. 4.2). The intermediaries seemed to exist at all levels of the increasingly general taxa. The bat, for example, is an intermediary between the class of quadrupeds (four-footed animals) and the class of birds (fig. 4.3). The ostrich forms a similar bridge between quadrupeds and birds, but it has more birdlike traits (sans flight). Quadrupeds are linked to fish through the intermediaries walruses and seals, while birds are linked to fish through the intermediary penguin. Buffon claimed that our carvings of nature into genus, order, and class were not reflections of the actual divisions in nature. Nature was a continuum without absolute divisions, and the taxonomic categories existed only in the human mind.

Given this view of nature as seamless, it is probably not surprising to discover that Buffon leaned heavily toward the kind of skepticism that we already encountered in John Ray. Buffon was squarely in the empiricist tradition, which believed that all knowledge comes through the human senses. And like the British empiricists Locke and Ray, with whom Buffon was quite familiar, this French skeptic understood that our senses could only deliver up the appearances of things. We do not have access to the God's-eye view of reality; we only have access to the subjective reality that our senses mediate for us. This theory of knowledge, together with a metaphysical conviction of nature's continuity, led Buffon to ridicule the search for a natural classification of essences. Appearances, or what the Greeks called phenomena, were the only things we could build our sciences upon. In other words, Buffon shared Ray's view that our human minds (mediated by the senses) were probably not isomorphic with the Deity's (unmediated) mind. But this orientation did not lead Buffon or other like minds to throw in the towel on scientific truth. They simply scaled back the hubristic version of natural science, discarding the we're-reading-God's-mind model of science. This form of skepticism, which acknowledged the futility of doing divine psychology and got on with the human uses of scientific inquiry, is generally called phenomenalism.

This emphasis on the phenomena, the appearances of things, was very disturbing to many naturalists of the period because they were afraid that one could not construct a proper science on something so fickle and fluctuating as the senses. Surely, they argued, science is about laws of nature, and laws are exceptionless, objective, universal truths. How can there be natural laws based on nothing but subjective sights, tastes, touches, and smells? The Linneans believed that to arrive at the essence of a species meant that one had finally arrived at the bedrock, the unchanging reality that allowed for objective lawful predictions. Buffon thought that this was sanctimonious twaddle and claimed that humans just clump their repetitive sensations into generalizations and call them "laws" or "essences" or "classes" or "orders."

For the system lovers, nature came packaged in neat rational bundles or norms, and variations from those norms were thought to be

Figure 4.1. Drawing of a "Water Shrew" from Buffon's *Histoire naturelle, générale et particulière.* (FROM DOVER PICTORIAL ARCHIVE SERIES, 1993)

Figure 4.2. Drawing of a "Suckling Pig" from Buffon's *Histoire naturelle, générale et particulière.* (FROM DOVER PICTORIAL ARCHIVE SERIES, 1993)

Figure 4.3. Drawing of a "Horseshoe Bat" from Buffon's *Histoire naturelle, générale et particulière.* (FROM DOVER PICTORIAL ARCHIVE SERIES, 1993)

rare freaks of little consequence. Animal and plant variations were just occasional obstacles to the otherwise smooth process of classification, random square pegs that did not fit nicely into the round holes. For Buffon, on the other hand, nature was variation through and through. There were no truly round pegs. Taxonomists simply ignored the subtle variations and forced different-shaped pegs into holes of their own invention. Buffon says of nature, "It is necessary to see nothing as impossible, to expect anything, and to suppose that all that can be is. Ambiguous species, irregular productions, anomalous beings will from now on cease to astonish us." Nature does not reveal some rational design or order; we ourselves impose order where none exists. We cannot detect the purposes of the Deity by studying nature. We need to give up the goal of trying to see nature on its own terms or on God's terms and just admit to ourselves that we are destined always to see it in our own human terms.

From this radical position, Buffon argues that the true natural system of classification will be the categorization of animals and plants according to their relative utility to mankind. This may not be nature's true structure, but it's the order of things most relevant and natural *to us*. Since the subjective cannot be expunged from science, Buffon is unapologetic about an anthropocentric taxonomy.

Once we understand Buffon's claim that nature is continuous rather than discrete, we do not suddenly start to see animals blending together—we do not suddenly lose our divisions of species and genus. The world continues to appear structured and classified, but that appearance is an inevitable by-product of the fact that we are human beings—the kinds of creatures who perceive in certain color spectra, who perceive atomic agitations as motionless solids, who have to eat certain kinds of animals and plants (and avoid others), who can use certain animal hides for clothing, and so forth.

The most natural way to classify animals, according to Buffon, would be by their use to man, so "those that are the most necessary to him, and most useful, will occupy the first rank; for example, he will show a preference to the horse, dog, ox, etc. in the animal order." Next he will take stock of wild animals that live in his own land, and then move on to the animals of foreign lands. This may seem bizarre, but Buffon asks: "Is it not better to arrange objects in the order and position where they are normally found, not only in a treatise on Natural History, but even in a painting or anywhere else, rather than to force them together because of a supposition?"

Linneans, according to Buffon, had started with a useful system but eventually deified the system itself and continued to extend it and worship it as an autonomous thing. Even the giddy obsession of the binomial nomenclature got under Buffon's skin. He argued for the use of common names rather than Latin technical language: "Why make up jargon and phrases when one can speak clearly, by pronouncing only one simple name?" And to obtain a universality to the terms, he argued, we should all agree to use the common name that the indigenous people use to name their local creature. Native Americans, for example, called the "muskrat of Canada" the ondatra, so Europeans should call it the ondatra also (fig. 4.4).

The last thing to note about Buffon's natural history is that his skepticism had an important boundary. As we've seen, he believed that most higher classification categories were useful fictions that human beings had made but then forgotten that they had made. This was not true of the species level of classification. Species classes were real divisions in nature, and we had an objective test for establishing these divisions. If you bred two animals and they had no viable offspring or the offspring was itself infertile, then this proved that the animals were of different species because their union could not contribute to the continuance of the species. Likewise, if even the strangest-looking pairing, such as a poodle and a hound, could produce fertile offspring, then their species solidarity was empirically established. Here was an experimental criterion, not a let's-look-at-their-shapes-and-colors criterion, that allowed even hardened skeptics the confidence to say something definitive about nature's categories. It was not always possible to try this experiment (getting

Figure 4.4. Drawing of an "Ondatra" from Buffon's *Histoire naturelle, générale et particulière*. (FROM DOVER PICTORIAL ARCHIVE SERIES, 1993)

diverse animals to have sex with each other before artificial insemination must have been a frustrating research program), but at least a workable principle could be agreed upon. This was an important moment in the development of classification and display because it set the stage for a different sort of organizing agenda. After Buffon had showed the futility of system for system's sake, naturalists started to move away from just morphological methods of classification. Looking for more productive avenues than Buffon's "utility" criterion, naturalists began to move from the appearances to the engine-room interiors of organisms.

Before we go on, it is interesting to speculate on a Buffonian museum display. One can imagine two different microcosmic rooms. The first could be labeled "The Real Nature" and the second "Our Nature." The first room would be like a return to the early curiosity cabinets, with specimens disarranged in a helter-skelter, unintelligible manner. The exception to this chaos would be the groupings of some species members together with each other, since species categories are real and can be established by experiment. But there would be no hierarchy of specimens in this room. The second room, devoted to the way in which nature prioritizes itself to us, would be arranged quite differently. Here one can imagine two human specimens (male and female) at the center of the room, probably skeletons or dry mounts. Emanating from these central figures would be concentric circles of specimens. The arrangement of flora and fauna would be according to their respective uses to humankind. Domestic animals and plants that provide food and clothing, such as cows, pigs, potatoes, hemp, and such, would be very close in proximity to the center. Beasts of burden would make up another category of concentric specimens. Medicinal plants could be gathered into concentric tables according to their necessity to human health, and so on.

The somewhat humorous implication of this taxonomic system is that every geographically different human race would have a different order of taxa. For Buffon, the whale and the Arctic seal would be on a table very far from the central human figures, but if Inuit people were to construct the display, these animals would be moved into a position contiguous to the human figures, and horses and cows would become remote.

In summary, Linnaeus was the great systematizer of organic surfaces (morphological structures), and Buffon was the foil who demonstrated the futility of system for system's sake. Buffon's skepticism about the cataloging mania, together with his claim that reproduction was an important key to classification, led subsequent generations to move beyond mere data collection. Data alone, while continuously piling up through the efforts of amateur and professional naturalists, were relatively mute. Arranging and rearranging specimens in classificatory tables was growing tiresome. What needed answering was why the organisms had just these characters. It was time to look into the engine room.

Here is a brief illustration, an analogy, of the way in which "dumb data" can nag and beg for explanation. While I was reading the 1831 catalogs of John Hunter's collection back in England, I ran across the following entry in the chapter titled "The Preparations of Monsters and Malformed Parts, in Spirit, and in a Dried State." A brief description and narrative accompany a specimen donated by one Nathaniel Highmore, Esq. The specimen description reads: "A female monstrous fetus found in the abdomen of Thomas Lane, a lad between fifteen and sixteen years of age, at Sherborne, in Dorsetshire, June 9th, 1814." This seemed bizarre, even for the Hunterian collection. Utterly confused, I jumped straight ahead to the diary-narrative of Highmore's postmortem examination for some answers.

Highmore's running commentary explains, "On dividing the parietes of the abdomen and exposing its viscera, a large tumour, of an irregular but somewhat oval form, presented itself." Finding that the boy's intestine formed a sack around the tumor, Highmore goes on to report that "[t]he spleen, which had previously been suspected to be the viscus diseased, was lying behind the tumour, appeared compressed by it, and was inflamed at its lower edge." After a thorough examination of the liver, kidneys, ureters, and urinary bladder, Highmore continued the autopsy. "As the tumour appeared to be the only diseased part, and was evidently contained within the intestine, with a free communication between it, the duodenum, and the jejunum, I proceeded to remove it from the body." It then becomes clear in the diary that Highmore was the sort of man who brought his work home at night.

> I conveyed the tumour to my house, and found its weight to be four pounds and a half: and, whilst in the act of placing it in a vessel where it would remain during the night, my fore-finger accidentally slipped around the curve of the arm, at the elbow joint, and first gave me the idea of its partaking of an animal form. At this circumstance I felt greatly astonished; and the more so, when, on my endeavouring to trace, through the sac, the extremity of the limb, I could count five digitals!

When I originally reached this point in the narrative, a point where even the clinical Highmore is compelled to use an exclamation mark, I could hardly contain my curiosity. When I turned the page in order to learn the solution to the mystery, I found only this abrupt closing line: "I determined on taking the fetus to London; where, having submitted it to the inspection of my medical friends, I deposited it in the Museum of the Royal College of Surgeons." That's it. That's all. No other information included, except that he did, indeed, recheck to confirm that Thomas was a true male. He was. The catalog, as if mocking my frustration, quietly moves on to a completely unrelated specimen.

One feels a nagging sense that something important is missing from the description of the specimen. Would one's mind become satisfied, or would the provocativeness of the specimen go away, if one was told of two more cases of fetal monsters growing in odd places inside male patients? Doubtful. I read on through the Hunterian catalog, finding other such specimens, but each account ended more abruptly than the first. What was needed was not more and more of these disturbing data; whether it was three such cases or three thousand, the illumination was equally truant. What was needed was a different mode of analysis altogether, a causal mode rather than a descriptive mode. How did the damn thing get there?

We'll return to this case later. For now, the frustrating experience with this example of provocative but mute data (the monstrous fetus) serves as a nice analogy for this particular phase in European natural history. Collecting, describing, and classifying organisms was no longer enough. The surface analysis had been exhausted. The burning questions of why and how, even in remarkable cases such as this one, had been unanswered, but now that began to change. Why did certain animals appear so similar when their environments were so different? How were all the forms of life related to each other? Why did embryological development seem to reveal additional morphological similarities between kinds of organisms? Why do totally different traits in different species all seem to fulfill the same organic function? Is this proliferation economical?

Two major trends emerge from this post-Buffonian period. The two seminal trends are actually just different expressions of the same impulse to penetrate the veil of appearances. Some naturalists began to pursue the causes of the individual organism (and the species), while others began to examine the causes of the whole organic chain. These approaches were physiological and evolutionary, respectively. By turning to physiology, naturalists such as Cuvier and Hunter attempted to unravel the causal pathways that led to the anatomical structures so beautifully cataloged by the system lovers. By introducing transformation theories, naturalists such as Lamarck attempted to give a causal foundation for the entire series of organic forms. One group was relatively local in perspective, while the other group was global, but both were moving from a descriptive paradigm to a causal one.

While standing before Cuvier's natural history cabinets, I thought about Ralph Waldo Emerson. In 1833 Emerson had toured these same display cabinets and invoked the experience in his essay *Nature*. In this passage we find a poetic expression of this causality phase in natural history. Nature cabinets, according to Emerson, allow for rational contemplation. Rational contemplation of nature's forms leads to something beneath the surface appearances.

> The first effort of thought tends to relax [the] despotism of the senses which binds us to nature as if we were a part of it, and shows nature aloof, and, as it were, afloat. Until this higher agency intervened, the animal eye sees, with wonderful accuracy, sharp outlines and colored surfaces. When the eye of Reason opens, to outline and surface are at once added grace and expression. These proceed from imagination and affection, and abate somewhat of the angular distinctness of objects. If the Reason be stimulated to more earnest vision, outlines and surfaces become transparent, and are no longer seen; causes and spirits are seen through them.

We'll discover that in their own separate ways, Lamarck, Hunter, and Cuvier were all trying earnestly to see these causes through the transparent surfaces.

Jean-Baptiste Lamarck, the brilliant French materialist, broke definitively with the long natural history tradition of theism. Unlike Linnaeus's pious belief that natural philosophy revealed God's role, Lamarck argued for a self-justifying cosmos. For Lamarck and other atheistically minded philosophes, matter itself, without God's interaction, could be the source of its own explanation. Nature, for Lamarck, was its own end.

European attitudes toward God and nature were conflicted during the early nineteenth century. On one hand, the Newtonian craftsman God seemed to be implied by the lawful processes of the nature machine, but on the other hand, the gaps in knowledge (where God was presumed to do his work) were steadily filling in without any need for spiritual causes. Everywhere you looked, the machine of nature was just making itself. Maybe God's ultimate purpose and justification for the cosmos truly existed out there, but increasingly God was neither helpful nor necessary for the explanation of natural events.

Interestingly enough, the paradoxical tension between nature as a produced artifact and nature as its own end was not easily reconciled in the European mind. It was partly this tension that led Romantics of the nineteenth century to shift away from the idea of nature as a machine and toward the idea of nature as an organism. If we think of the whole cosmos as alive, as many Romantics did, it becomes easier to see nature as its own end. Organisms are more obviously ends in themselves than are machines. For philosophically minded naturalists of the later Romantic period, the cosmos ceases to be an artifact and instead returns to the more Greek notion of a cosmic organism.

One can see how this attempt to construct an internally coherent science of organisms is another result of the phenomenalist movement, which doggedly persevered in research despite the skeptical impasses. Natural history went from trying to read God's blueprint to pretending we were reading God's blueprint to admitting that nothing interesting was being added by the invocation of the God hypothesis. Laws of nature make sense even when there's no lawmaker. In this respect, Darwin was only finishing a revolution that started during this causal phase of natural history.

Most of biology (a word that Lamarck invented) was becoming increasingly specialized during this period, but Lamarck was old-fashioned in his scope. His atheism was not old-fashioned, but his attempt to explain the totality of nature was. The Great Chain of Being was a generally recognized pattern of nature at the time. There seemed to be an ascending ladder of animals from simple to complex, with each link of the chain resembling (but also departing from) the links above and below it. The great apes, for example, resembled the less sophisticated monkeys below them, but also the human beings above them. And within the major divisions, further splits revealed the seamless chain of relations, because chimps seemed higher than gorillas, which seemed higher than gibbons, but orangs seemed higher than chimps, and so on. Lamarck took this old European assumption of a chain of being and reinterpreted it as a temporal sequence. Things were higher and lower on the chain not because God had laid it out that way, but because simpler organisms evolve into more complex organisms over time.

Following Buffon's lead, but radicalizing it further with the addition of transformation, Lamarck argued that the divisions of taxonomy were only fictional breaks in a truly continuous nature. Nature was a dynamo of constant change, and in his 1801 book *Système des animaux sans vertèbres* he arranged animals in a sequence that depicted the general trend of this constant change. In his 1809 book *Philosophie zoologique* Lamarck articulated his two major laws of this evolutionary process. The first was that all living creatures had an inherent trend toward the next highest link in the chain of complexity. And the second law was the "inheritance of acquired traits," an idea that has been a favorite punching bag for twentieth-century neo-Darwinians.

While the inheritance of acquired traits turned out to be the theoretical equivalent of the Edsel, it has to be admitted that it was not a bad first attempt to link organisms to their environments without appeal to God. The major puzzle in any evolutionary theory is how an organism's characters fit so beautifully with its specific environment. If an animal's actions in this lifetime change its body in ways that can be passed down to its offspring (the classic example is of the giraffe, whose neck-wrenching foraging behavior results in a longer neck, which is inherited by its offspring), then you can theoretically link an animal's changing body to the animal's changing environment. The theory provides an adaptation-tracking mechanism. Of course, the major stumbling block to the theory of acquired traits is that it's totally false.

If offspring inherited acquired traits, it certainly would speed up the evolutionary process—organisms wouldn't have to wait for changes to occur through mutation of the genome. If I wanted my child to be born stronger, I'd pump iron during my lifetime; or if I wanted her to be born smarter, my wife and I would spend more evenings curled up with Euclid before we conceived. Of course, this isn't how biology works, but many mice had to lose their tails to prove it.

In 1904 August Weismann, a German zoologist and embryologist, published his experiments in a book called *The Evolution Theory*. He analyzed the Lamarckian theory of acquired-trait inheritance, a theory that most naturalists (including Darwin) took seriously in the late nineteenth century, and found it lacking. He cut off the tails of a male and female mouse and bred them. Next he took the offspring of that union and repeated the process. Weismann, in a laboratory rodent binge, carried the experiment through twenty-two generations, and the tails of the offspring never got any shorter than those of other mice.

Now one might object, and some did, that Weismann did not do the experiment well enough or long enough or carefully enough (for despite our schoolhouse version of science, real scientific disputes are rarely solved by experiment alone, since opponents can always interpret the experiment differently). But suffice it to say that the onset of modern genetics and our eventual grasp of DNA transfer eventually killed Lamarckism, because no mechanism could be found for moving acquired traits back into the gamete's genome (genotype-to-phenotype expression is a one-way street). But the failure of Lamarck's mechanism of evolution shouldn't blind us to his bold and visionary commitment to evolution itself. And this shows up clearly in his museum cabinets.

At the same time that Cuvier was curating the zoological cabinets of the Muséum national d'histoire naturelle, his colleague Lamarck was in charge of the invertebrate collections. And here inside the Galerie d'anatomie comparée I noticed one of Lamarck's cabinets, sitting lonely and overwhelmed by the dominating presence of Cuvier's armoires. This case of Lamarck's specimens is only a contemporary remnant of a once-formidable collection, a collection now scattered and decontextualized from its original presentation plan. If the history of science is unkind to you, then not only do your books go unread, but your stuff gets irreverently shuffled around as well.

I studied the cabinet for a while. It was about seven feet tall and three feet wide, with a glass casing on all sides. The case was old, but it did not appear to go back to earlier than the 1890s. There was a sculpted but simple metal frame around the cabinet, and the lower quarter consisted of metallic specimen drawers. The cabinet made a stark comparison with the nearby metalwork of the balcony handrails. These ornate Art Deco forms were heavily stylized plant shapes that repeated an explosion of iron tendrils every two feet. They circled playfully above the otherwise very serious room.

Inside the cabinet was a combination of corals, fossil worms, and shells, mostly of mollusks. There was a little white card with the neatly penned label "Espèces décrites par Lamarck." A larger and more formal note mounted on the side of the case read "Exposition presentée à l'occasion du 250ème anniversaire de la naissance de Lamarck (1744–1829), naturaliste français, père de l'évolutionnisme." The curator had placed the specimens in a pattern that was true to Lamarck's original presentation strategy, from the simple to the complex.

If one studies the morphology of some invertebrates, one can't help but think of them as simpler, less perfected versions of vertebrates. We know now in the twentieth century that this is just anthropomorphic (or vertebratemorphic) thinking. We know now that the onward-upward-progress sequence of life from less to more complex is just one small thread of the wider story of life, a story that moves in many directions at once and even runs in place, but we cannot fault Lamarck for getting excited about the invertebrate patterns. Invertebrates can be placed into a relatively tidy hierarchy of forms based on the way they look and function, and Lamarck was digging hundreds of these things out of the geologically stratified Paris basin.

Invertebrates, for example, display some of the following characters, and once these are appreciated, it is easy to see the logic behind Lamarck's ordering system. The phylum Coelenterata, which contains jellyfish, sea anemones, and coral, is characterized by the "primitive" trait of radial symmetry (all sides mirror each other) and a primitive germ-layer construction (they possess ectoderm, mesoderm, and endoderm tissues). The phylum Platyhelminthes (flatworms, such as flukes and tapeworms) seems to be slightly higher than Coelenterata because flatworms are bilaterally symmetrical (like us) rather than radially symmetric, and they possess primitive nervous and digestive systems. Still, these invertebrates seem trumped by the Aschelminthes (roundworms), who have a complete digestive tract, with a mouth and an anus. But the roundworms still lack a proper circulatory and respiratory system (of course, to talk of one animal lacking another animal's equipment is already loaded with a value judgment that doesn't really make sense in Darwinian evolution). So the worms lose to the mollusks (oysters and snails), those favorite sons equipped with well-developed (i.e., more like our) digestive, circulatory, respiratory, and nervous systems. But the worms make a comeback in the phylum Annelida (segmented worms, such as earthworms), unabashedly lording their segmented superiority over the mollusks. Not only are the earthworms in possession of a closed circulatory system, but they have a very complex mesoderm-divided body cavity. And segmentation is a sure sign of superiority, since vertebrates possess considerable segmentation. Unfortunately for the earthworms, who desperately want to be more like us (the vertebrates), another phylum outstrips their noble effort. The invertebrates known as sea squirts and sea lancelets are almost in the vertebrate club because they possess a supportive structure called a notochord that runs through their long bodies (in truth, these animals were not placed very high in Lamarck's scale of being because of different anatomical emphases). My favorite of these chordates is the aptly named slime hag, a worm-shaped deepwater scavenger whose defense mechanism is to tangle its enemy in a soup of slime that it ejects from glands all over its body. It's this kind of weirdness that makes one wonder whether Aristotle's bonasus is really out there somewhere.

It should be clear even from my thumbnail version of Lamarckian progress evolution that it was at least reasonable during the early nineteenth century to classify organisms in this hierarchic way. Unfortunately, just because we can put invertebrates into a sequential order of complexity does not mean that this sequence tells us the actual history of life. We can also order together all red objects that are smaller than a house cat, but we have not captured some natural taxonomic group by doing so, nor have we isolated a real causal connection between the things ordered. What we recognize to be higher and lower is determined by which characters or traits we choose to analyze.

Lamarck thought of evolution as the engine room that churned out the diversity of kinds and types that our classification systems tried to mirror. While he was quite wrong about how the engines functioned, he was essentially correct in trying to tie taxonomy to a causal evolutionary story. But Cuvier could not abide this feature of his colleague's research.

Cuvier and Lamarck agreed that the fossil species they both were digging out of the geological strata did not represent currently living forms. Extinction was a new concept at this time, but the growing fossil record quickly made the phenomenon undeniable. Lamarck felt comfortable speculating, in light of this new concept of species death, that evolution gave birth to new life forms; evolution caused organisms to have their specific characters and indeed their very existence. But Cuvier was agnostic on this point. He believed that geological catastrophes brought about extinction, but on the issue of species creation he offered no positive theory (though he did rant vociferously against Lamarck and other transformists). He remained on the skeptical fence about the creation of life forms because he felt that we could not have the requisite experience to resolve such matters. Cuvier saw his colleagues Lamarck and Etienne Geoffroy Saint-Hilaire as overly speculative and easily given to flights of imagination when they should be down in the trenches with him doing empirical research. In the "Discours préliminaire" of Cuvier's *Recherches sur les ossemens fossiles de quadrupèdes*, he jabs at his evolution-minded colleagues. He begins by saying that some contemporary minds (minds that are too "liberal") are pursuing the strange idea that "everything was fluid in origin; that the fluid generated animals that at first were very simple, such as monads or other microscopic species of infusoria; that in the course of time and by taking up diverse habits, the races of these animals became more complex, and diversified to the point where we see them today." Then, in a continuation of this passage, a continuation that was edited out of the published version, Cuvier gets bitingly funny in his characterization of Lamarckism.

> [T]he habit of chewing, for example, resulted at the end of a few centuries in giving them teeth; that the habit of walking

> gave them legs; ducks by dint of diving became pikes; pikes by dint of happening upon dry land changed into ducks; hens searching for food at the water's edge, and striving not to get their thighs wet, succeeded so well in elongating their legs that they became herons or storks. Thus took form by degrees those hundred thousand diverse races, the classification of which so cruelly embarrasses the unfortunate race that habit has changed into naturalists.

Contained within this amusing roast of his contemporaries is the true kernel of Cuvier's major fear. While he was skeptical and agnostic about a cosmic causal analysis of natural history (Lamarck's evolution), he was in fact utterly dedicated to discovering the lawful and causal engine room for each individual species. If evolution was true, if Buffon and Lamarck were right about the inherent fuzziness of all taxa, then, Cuvier reasoned, a real science of natural history would be impossible, because there would be no fixed subject matter, only fugitive forms (that is why classification science would be so "cruelly embarrassing"). Natural history could be made into a law-abiding discipline not by linking the species together in some great chain, but by analyzing the physiology of the organisms and seeing how the taxonomic distinctions recognized by the systematists all flow from the laws of physiology. In this way Cuvier, like some of his Enlightenment predecessors, wanted to put the shaky science of natural history on the solid footing enjoyed by the physical sciences. In physics there were exceptionless laws (e.g., F = M/A) that allowed for successful predictions about future phenomena, but in biology (like history) everything seemed to be an exception rather than a rule.

Both Cuvier and Hunter, working independently (though a case might be made for Hunter's priority), reoriented natural history away from the structures and toward the functions of organisms. Neither of them was interested in articulating a descriptive table of creatures. Both of them were more interested in articulating a predictive set of laws that had to rest on a foundation of causal principles. Thus Hunter was experimental rather than an exclusive recorder of field notes about animal behaviors.

We are finally ready, after this short history of taxonomy, to appreciate Cuvier's cabinets. Among other things, we now have some appreciation of the concept of biological "characters," some familiarity with the quest for a natural system of classification, some acquaintance with the skeptical and phenomenalist traditions, and some understanding of natural history's desire to cross over from descriptive science to law-based predictive science.

One of the first things that one notices as one wanders through the Galerie d'anatomie comparée is that Cuvier's preservation techniques were not up to the level of Hunter's. Perhaps this is a reflection not on the two men but on the generations of subsequent caretakers. Whatever the case, the wet preps here are all much murkier (the alcohol needs freshening), and the skeleton specimens could use a couple of hours in the flesh-eating dermestid tanks (something to strip off the brownish chunks of flesh and sinew).

Cuvier began as an assistant in animal anatomy in 1795, but he was so talented and ambitious that by 1802 he was named chair of comparative anatomy. It was at this time that he began to construct his collection from the more than two thousand specimens he dredged out of various basements on the Jardin's grounds. When Cuvier died, in 1832, he had built up his cabinet to 16,665 specimens. In the Galerie now, it is somewhat difficult to determine which cabinets are original Cuvier groupings and which are the product of later curating. Fortunately, the excellent historian of science Phillip Sloan has painstakingly researched some of these changes through early catalog descriptions and letters. Much of my understanding of the original Cuvierian and Hunterian layouts is indebted to his fine scholarship.

As one continues to wander, the next important thing that one notices is that Cuvier's armoires (unlike Hunter's) are not populated by wildly different classes of specimens. In just one of Hunter's cabinets you can find sap, blood, muscle fibers, shells, horns, joints, and feet—these are grouped under the title "Parts Employed in Motion." Or in another Hunterian cabinet entitled "Nourishment, and Protection, Afforded by the Mother to Her Young," one can find the following specimens side by side: milk glands and nipples, crop glands of birds, young animals attached to the tail of the mother, young in pouches, young on the mother's back, and young animals protected in nests. Cuvier's cabinets, on the other hand, are much less syncretic (fig. 4.5).

Among the first few display cases that I encountered were ones focused on the sensory equipment, ears, noses, eyes, and such. But while an armoire may contain several different species' olfactory equipment (a wet-prep elephant trunk cutaway that looks as though it is floating in clear jelly, a cat skull sinus cutaway, a human nose and sinus prep, etc.), the cabinet never veers from vertebrate specimens—and frequently the cabinets admit only mammals or fish or reptiles, unmixed. Vertebrates and invertebrates are never in the same cases. I found the same general trend in the displays of nervous systems, circulation systems, digestion, respiration, and so on.

Figure 4.5. A Cuvierian armoire of primates. No *embranchement* divisions are crossed in this sort of display practice. *Galerie d'anatomie comparée*, Paris. (PHOTOGRAPH © RICHARD ROSS)

As I was contemplating a particularly nasty specimen of a chimpanzee's ear, floating complete with several inches of surrounding hairy scalp, I reflected on Cuvier's theory of higher taxa. Cuvier argued that major taxonomic divisions emerged when we studied animal functions of nutrition (alimentary canal) and motion (the required functions of the nervous system, circulation, and respiration). The four major divisions that he derived from this functional analysis were called the four *embranchements* (branches): Vertebrata (birds, mammals, etc.), Mollusca (such as oysters and snails—Cuvier began poaching on Lamarck's territory by arranging these invertebrates according to his own system), Articulata (e.g., insects, segmented worms), and Radiata (such as the starfish and jellyfish). Notice how these *embranchements* fracture the continuity of Lamarck's invertebrate/vertebrate ladder. That is no accident.

The *embranchement* method of division did indeed recognize important functional/structural differences between formerly lumped-together classes (such as insects and coral) and for this reason the system had intrinsic virtues. But for Cuvier it also had extrinsic virtues. With the *embranchements* established as completely different and unbridgeable classes, the evolutionary chain of gradations could find no traction. The transformists, including Lamarck, were trying to show continuity in their theoretical taxonomy and in their practical museum displays, but Cuvier's taxonomy stressed discontinuity between classes, and his museum displays did the same.

The function-based method of dividing the *embranchements* was used in all of Cuvier's other taxonomic work, and he came to summarize this approach with two principles: the "conditions of existence" and the "subordination of characters." The naturalist could best understand how and why an animal was constructed by thinking about both general and particular animal functions. The shorthand way of talking about these functions, and emphasizing their centrality, was to refer to them as the "conditions of existence" (old-school naturalists called these "final causes," after Aristotle, but that term had become confused by religious agendas). According to Cuvier, focusing on the conditions of existence finally made taxonomy scientific (it brought us as close to a natural classification as we are ever going to get), clarified the work of paleontologists, and gave anatomy a rational basis. So, in a way, Cuvier never really gave up on the Linnean program; he simply reversed it. For Cuvier, one does not first attempt to set up a cosmic table of organisms by observing and classifying. Such a table can come only *after* the causal principles or laws have been discovered. The conditions of existence are those causal principles that will make a more principled and coherent taxonomy.

In his "Discours préliminaire" Cuvier explains how all this fits together. A growing problem for the naturalist at this time was how to make heads or tails of the fossil parts coming out of the ground. Finding complete fossil skeletons was incredibly rare (bits and pieces were the rule), but even in that rare case, the skeleton alone was relatively unhelpful for taxonomic identification because most previous classifying, generally speaking, had focused on the surfaces—skin pattern, coloration, and so on—that are long gone in the case of fossils. But now the principles of conditions of existence and subordination of characters save the day. Using these principles, Cuvier can detect a "correlation of forms in organized beings, by means of which each kind of being could be recognized, at a pinch, from any fragment of any of its parts."

The conditions of existence principle will first send us looking for the manner of an animal's nutrition (what and how it eats in order to have an existence) or some other major physiological system. This fundamental character (a new spin on the old essential character) subordinates other characters beneath it in a predictable and logical fashion. Cuvier explains that "if the intestines of an animal are organized in such a way as to digest only flesh—and fresh flesh—it is also necessary that the jaws be constructed for devouring prey; the claws for seizing and tearing it; the teeth, for cutting and dividing its flesh; the entire system of its locomotive organs, for pursuing and catching it; its sense-organs, for detecting it from afar; and it is even necessary that nature should have placed in its brain the instinct necessary for knowing how to hide itself and set traps for its victims." Cuvier calls this description the "general conditions of existence" for the carnivorous regime. This mode of thinking is a major work of genius on Cuvier's part (we in the late twentieth century are so familiar with this mode of analysis that we forget its impressiveness). Cuvier forges a way of thinking that treats the animal itself like a logical syllogism—but instead of propositions entailing each other, it is biological characters.

In discussing this same carnivore syllogism, Cuvier points out that the conditions of existence can be particular as well as general: "[W]ithin these general conditions there exist particular conditions, relative to the size, species, and habitat of the prey to which the animal is adapted; and each of these particular conditions results in some detailed circumstances in the forms that result from the general conditions. Thus it is not only the class that finds expression in the form of each part, but the order, the genus, and even the species." In this way, all the taxonomic categories find objective scientific grounding; that is to say, law-like certainty.

So the particular problem of digging up a confusing fossil is resolved by using the same logic of physiological anatomy that guides all of Cuvier's work. The causal engine room that he discovers reveals that *function* causes all organic structures. He states, "Every organized being forms a whole, a unique and closed system, in which all parts correspond mutually, and contribute to the same definitive action by reciprocal reaction. None of its parts can change without the others changing too; and consequently each of them, taken separately, indicates and gives all others."

Cuvier's conviction that each animal has a logical coherence to it, a coherence that ties its parts to each other and to the environment, allows him to dismiss evolution and also affirm the scientific certainty of biology. Evolution could not be occurring, he reasoned, because even the slightest change in one of the organism's body parts would disrupt the complex harmony of structures and functions. An organism is like a house of cards: remove or alter one part, and the whole thing collapses on itself. And in addition to this ontological feature of Cuvier's system, the epistemic flip side maintains that the invariable connections of each character allowed for reliable inferences about nature's plan. He's quite clear about his own contribution, for example, in his 1796 analysis of the *Megatherium* skeleton of Paraguay (fig. 4.6). After using his twofold principles to establish that the animal "should be placed between the sloths and the armadillos, since it combines the shape of the head of the former with the dentition of the latter," he states that his whole analysis is a "very powerful proof of the invariable laws of the subordination of characters, and of the justice of the consequences that have been deduced from them, for the classification of organisms." In one masterly economic move, Cuvier cuts off the possibility of evolution with his metaphysical physiology, levels Lamarck's ladder of progress by making each *embranchement* equally perfect in its respective environment, and sets classification on the nonarbitrary foundation of physiological laws. All of this guided his curatorial decisions in the museum.

The most telling example of subtext in Cuvier's cabinets is probably manifested in his placement of fossil and living (*actuel*) life forms. A vestige of Cuvier's philosophy remains in the Galerie because most of the extinct life forms are housed on the second floor, above the main room. I had to gain access to them via a dark stairwell beneath a tiny balcony, where a sleepy-eyed security guard stood watch. But during Cuvier's day the fossil bones were not housed in this space at all. All fossils were separately displayed in another gallery in the Jardin des plantes. This is not because Cuvier was uninterested in fossil remains; on the contrary, he was absolutely instrumental in excavating and classifying the growing fossil record. But he was reluctant to display living and fossil forms in the same exhibit space, lest people associate older and recent forms (fig. 4.7). If people saw modern and ancient animals together, they might link them through the "heretical" idea of evolution; they might see them as genealogically related rather than just morphologically similar. The display rhetoric of the Galerie spoke against rather than for evolution.

When Cuvier visited Hunter's collection in 1818 he must have been both inspired and frustrated. For years the Frenchman had heard stories of a brilliant collection across the Channel, and in 1814 he had firsthand reports from his brother Fredrick and from Lamarck to whet his appetite further. What he saw when he finally made it to London was inspirational, because the specimens revealed Hunter to be a kindred spirit in his emphasis on physiology. But the collection must have been equally frustrating, because Hunter's catholic inclusion of very different families of specimens seemed to imply evolutionary inferences as well.

Cuvier was laboring in his collection to reveal how each animal was "related" to itself (conditions of existence and subordination of characters) but *not* related to other animals (no crossing of *embranchements*). Having to walk the hallways of the Royal Academy of Sciences with his evolutionary enemies Lamarck and Geoffroy

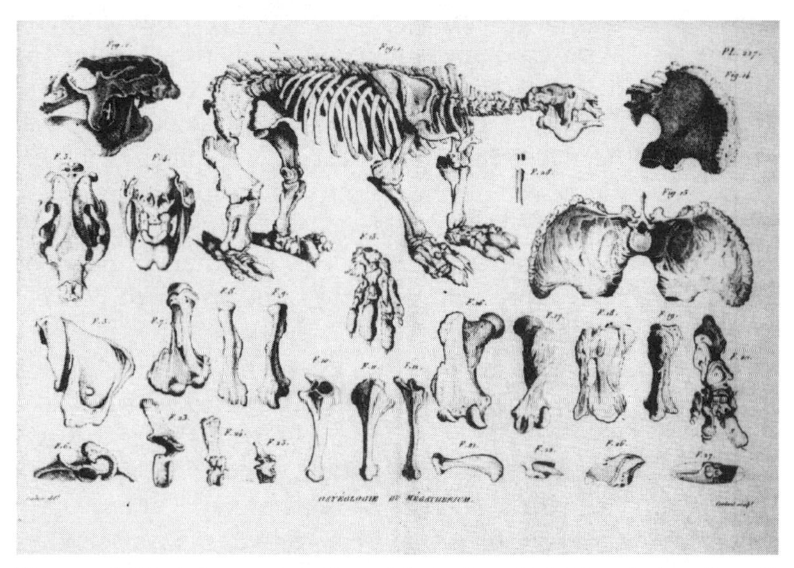

Figure 4.6. Cuvier's reconstruction of the extinct sloth *Megatherium*, from South America. *Récherches sur les ossements fossiles*, 1825. (FROM IMAGES OF SCIENCE, BY BRIAN J. FORD, © 1992 BY BRIAN J. FORD. USED BY PERMISSION OF THE BRITISH LIBRARY)

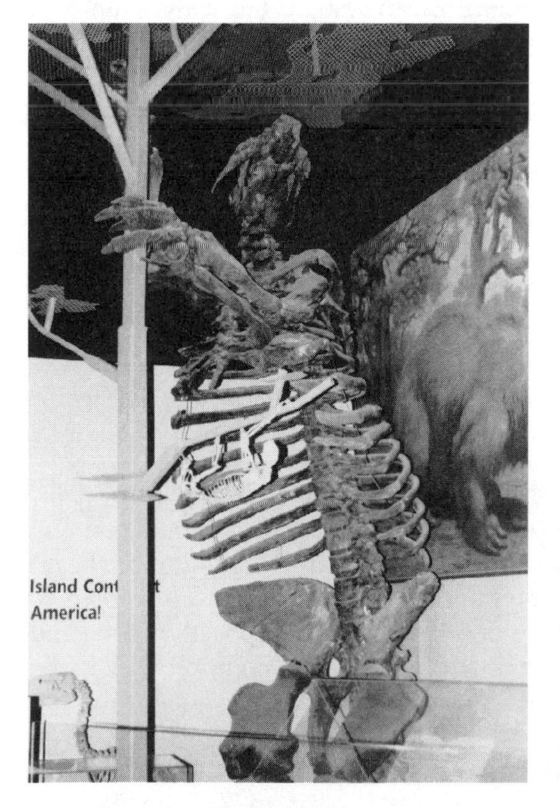

Figure 4.7. An extinct giant sloth (standing in background) posed with a small contemporary sloth (hanging from tree in foreground), at Chicago's Field Museum. This pairing of extinct and contemporary forms, in the same display, was strenuously avoided by Cuvier.

made Cuvier nervous and sensitive about his engine room of physiology. There was an obscure but tantalizing competitor (evolution) that Cuvier wanted to rub out. Hunter also wanted to reveal the physiological causes of the phenomena that previous naturalists encountered, but he never felt the threat of a competing metaphysics; consequently, he did not hesitate to group specimens across taxonomic divisions (mollusks and mammals together, for instance) or to display fossils and contemporary forms together. In point of fact, Hunter was even more impressed by, and driven to convey, functional physiology than Cuvier.

Hunter's intoxication with functional systems led him to see through all traditional morphology-based classification systems like so many transparent tissues of deception. Like Cuvier, he recognized the basic physiological functions of digestion, circulation, reproduction, and so on, but with syncretic vision; he could not see any importance to the surface divisions of vertebrates, invertebrates, articulates, and so on. Hunter even went so far as to display together a sea horse's tail and a plant's tendrils under the organizing principle "Methods of Support with Motion." This taxonomic category is obviously miles away from the Linnean system, but while Hunter is rather idiosyncratic here, we must acknowledge that this novel exhibit forces us to see and think about nature in a different way. Juxtaposing plant tendrils and sea horse tails gets us to think about new categories of function that we might term life's strategies. A museum arranged in this way produces a radically different emotional and cognitive response in the audience. The patrons of the earlier collections were struck by nature's (God's) prolific fertility of forms, but in Hunter's museum one witnesses how the diversity of forms is actually wedded to a rational economy of functions. As a result of this, Hunter's collection shows us a nature that is an active, productive force (*natura naturans*) rather than a passive product (*natura naturata*).

In a "Digestion" armoire, Hunter has two shelves devoted to "parts analogous to teeth in invertebrates," including parts of parasitic worms, cicadas, locusts, Roman snails, slugs, cuttlefishes, and squids; then "parts analogous to teeth in birds," including the beaks of vultures, woodpeckers, and puffins; then two shelves devoted to "true" teeth, including those of lions ("teeth composed of bone and enamel") and those of horses and elephants ("composed of ivory, enamel, and cementum"). Two things are fascinating about this kind of display.

The first compelling fact is that the sequence of specimens has something of the Lamarckian progress spirit to it. Starting with imperfect and analogous teeth, the display progresses through and culminates in mammalian teeth. This does not mean that Hunter was Lamarckian; indeed, Lamarck was only a relatively obscure novice when Hunter died. It also does not mean that Hunter was an evolutionist; he was simply a subscriber (like everyone else) to the very entrenched metaphysics of the Great Chain of Being. Nonetheless, this sequence, of simple to complex, could form associations in the minds of patrons—associations that made Cuvier bristle.

The second, and related, feature of Hunter's "Digestion" armoires (and all his other armoires) was the fact that he had abstracted dentition (and pseudo-dentition) entirely from the host organisms. The character was detached from its particular context and placed with other similarly excerpted traits. Hunter had such a comparative mind-set that no lumping of diverse structures seemed offensive to him, whereas Cuvier would always display a mammal's tooth, for example, with the other subordinate and dominant characters of mammalian digestion (never with invertebrates' eating equipment). In the 1822 DeLeuze catalog of Cuvier's cabinet, for example, the only comparisons of teeth (occurring in the third and fourth rooms of the second floor) occurred securely within the parameters of the vertebrate *embranchement*.

Cuvier's method of analysis, remember, is to relate the parts of an organism to each other and to specific environmental conditions of existence. So the lungs of whales, for example, have a real causal relationship with the other anatomical parts of whales (trachea, blowhole, circulation system, etc.) and with the needs and demands of its aquatic environment. But there is no causal relationship with other animals that have lungs. For Cuvier, there is parallelism of the whale's lung and the cow's lung, but nothing more. So homology did not reveal genealogy for Cuvier; rather, it revealed sameness of environment or sameness of function. Hunter and other comparison-minded naturalists, on the other hand, tended to group animal organs and functional systems into general categories that ignored the specifics of each species's related characters. In this way, animals in Hunter's collection were displayed more like pastiches than Cuvier's self-contained organic syllogisms.

Cuvier argued that the parts of organisms could not evolve because they were so intimately integrated with the survival systems, and change would undo the delicate organic balances. A heart's evolving of a new chamber, for example, would be as detrimental to the animal as the mid-recipe manipulation of ingredients would be to the success of a baked soufflé. His original arrangement conveyed this message to patrons by moving from full vertebrate specimens (skeletons and muscular models) on the ground floor to dissected viscera and physiological systems (wet preps and wax injections) on the second floor. As the patron ascended through the collection, he gained greater access to the hidden internal balance that showed each organ's necessary connection to every other organ. There was a unity to life, but for Cuvier that unity referred to the individual species, at least, and the individual *embranchement*, at most. In Hunter's museum, however, the unity of life extended, without jumps or gaps, from one end of the organic continuum to the other.

The Hunterian presentation of animal characters and physiological systems in abstraction leads to the idea that an organism is more like a Lego sculpture than a house of cards. Like a Lego sculpture, an animal is just an amalgam of the more fundamental parts; remove, add, or reposition a part, and the result is another kind of animal. Nature, for Hunter, seemed to have created all the different animal forms from a finite cache of similar materials and systems. He had discovered the cache and now set about displaying these underlying continuities beneath the apparent differences. Hunter argued, for example, that the human fang-tooth (canine) was just a less effectual version of the same structure and function found in a lion's dentition. The fang itself existed like a continuum; somewhat carnivorous animals (humans) on one end and supercarnivores (lions) on the other. When discussing the "Arrangement of Anatomical Preparations" in volume one of his *Essays and Observations*, Hunter explains, in a very revealing passage, that he is torn between practical and theoretical display techniques. With large animals he is pragmatically obliged to prepare and display the parts of the animal exclusively, but with small animals he is compelled to prepare and display the whole thing. In striving to reveal the similar systems of these different-sized animals (say, digestion or circulation), he realizes that the abstracted organs and the organs in situ will be potentially confusing to the patron who is familiar with traditional taxonomy. He admits that his display technique "breaks in upon the arrangement and classing of specimens: but it is somewhat like Nature Herself; one property belonging to more animals than one." And in the introduction to the *Essays* he explains: "We may observe that in Natural Things nothing stands alone; that everything in Nature has a relation to or connexion with some other natural production or productions; and that each is composed of parts common to most others but differently arranged."

Cuvier saw this idea of a promiscuous nature as threatening, because if animals were composites of interchangeable parts, then they might theoretically change over time. This would throw doubt upon Cuvier's idea that every part of every animal was "necessarily" fitted and dependent upon the other parts. Hunter's cabinets seemed to suggest that the only necessity occurred at the level of cross-species physiology (i.e., circulation, digestion, etc.), and all other specific characters were relatively contingent. At the outset of his *Essays and Observations* Hunter, while not proffering an explicitly evolutionary doctrine, nonetheless acknowledges that most of the plants and animals we encounter have probably changed significantly over human history (through either geological changes or domestication). According to Hunter, change is inherent in nature, and it is part of the naturalist's job, as much as epistemically possible, to track it through time.

This issue, of Cuvierian invariant nature and Hunterian variant nature, makes some sense out of Cuvier's and Hunter's respective emphases on anomalous specimens as well. Compared to Hunter's museum, Cuvier's display of teratological specimens ("monsters") was quite meager. The 1803 Cuvier catalog describes only a few human dwarf skeletons, and the 1822 catalog describes a few monsters contained in the very last room of the second floor. Cuvier conducted visitors through the elaborate presentations of nature's lawful and predictable formulations and concluded with a brief nod to the rare anomaly. Nature always gets things right; this is the confident metaphysical foundation on which to rest a legitimate science. Hunter, on the other hand, almost revels in the sundry ways that nature gets things wrong. The pathological, abnormal, and malformed specimens of Hunter's collection numbered over three thousand by the time of his death. Hunter's museum gives one the impression that nature is an inspired, experimental, but messy artist, whereas Cuvier's museum gives one the impression that nature is a tidy, detail-oriented bureaucrat.

We mentioned in chapter 2 that Hunter's emphasis on continuity, together with his exhibits of nature's trials and errors, may have had a formative role in Darwin's later commitment to transformation. If Cuvier had lived to witness this kind of museum effect, one suspects that he would have lamented it.

Cuvier's antievolutionary stance was not the result of naive religious piety. He simply believed that without strict, unchanging biological boundaries, biology would not be able to obtain the status of true science. We have this same bias in our contemporary attitudes about science, so we should not be too quick to judge. We talk about there being two kinds of science, soft and hard. And apart from the humorous gender implications of this distinction (implications wonderfully explored by feminist philosophers), a clear cultural inclination is manifested. Hard sciences (e.g., physics and chemistry) have been especially prized because traditionally they have been grounded in mathematics, and this makes their predictions and theories more universal in character. Mathematical truths, and by extension the truths of hard science, are true for all people, all places, and all times (indeed, in some cases even true for all bits of matter). Pretty impressive. Soft sciences (e.g., sociology, psychology, etc.), on the other hand, are mushy by comparison. They deal more with particulars than with exceptionless universals, and where they are willing to make predictions and general theories, they can only offer plausibility, trends, and probabilities. And during the nineteenth century biologists such as Cuvier were trying to rescue their discipline from the quagmire of soft science and place it in the camp of universalizing science. If the mathematically oriented sciences are so reliable because numbers don't change, then the biologically oriented sciences will be reliable because nature doesn't change (or so Cuvier believed).

Hunter's attitude toward the boundaries of nature were much closer to Buffon's than Cuvier's. Trying to make sure that every species or genus was carefully demarcated from other species (and displayed accordingly) was a waste of time, according to Hunter. Like Buffon, Hunter maintained that animal breeding reveals true *species* categories, but variety, genus, and class all shade into hopeless gradations that naturalists cannot begin to discriminate. I'm reminded of Cuvier's sarcastic suggestion that taxonomy would be a cruel frustration if nature was always changing: "Thus took form by degrees [in the course of evolution] those hundred thousand diverse races, the classification of which so cruelly embarrasses the unfortunate race that habit has changed into naturalists." Of course, Hunter's (and Buffon's) skepticism about classification is premised on nature's complexity rather than its transformation, but the general point is still clear. Hunter would not be bothered by Cuvier's criticism because, in light of the new physiology, classification by itself was ceasing to be an interesting research program altogether. Both Hunter and Cuvier were ushering in a new causal paradigm of natural history. But where Cuvier harked back to an Enlightenment faith in nature's invariant, machinelike dependability in order to ground his principles and laws, Hunter believed that the reliability of his emerging laws and principles stemmed directly from his experimental methodology.

Recall that from the sixteenth century up through the time of Linnaeus and Buffon, natural history and physiology were unrelated disciplines. Physiology was previously believed to be the purview of the physicists, and its research program consisted primarily of attempts to conceptualize organisms as machines (since that was the successful metaphor of the Newtonian/Cartesian era). But here in the late 1700s and early 1800s, Hunter and Cuvier were demonstrating that physiology had important implications for the study of natural history. Understanding the general functional systems (à la Hunter) and the specific functional systems (à la Cuvier) led to new taxonomies, new understandings of adaptation, and new tools for assessing the explosion of fossil forms. Hunter states that simple field observations will no longer suffice; physiological experimentation is now necessary. "In animals we must observe the natural operations of the animal; and, where opportunity does not serve us to observe Nature in her operations, we must put her in the way of yielding the means of observing those natural processes."

Hunter and Cuvier were conscious that they were moving beyond the surfaces of organisms to the engine room. Cuvier, in his "Discours préliminaire," points out that his theory of the organic syllogism (the conditions of existence and the subordination of characters) is based upon experimental observation. Given his new method, the smallest bone or structure can lead the smart naturalist to deduce every other bone. He declares: "I experimented several times with this method, on portions of known animals, before entirely placing my confidence in it for fossils; but it always succeeded so infallibly that I no longer have any doubt about the certainty of the results it has yielded." Likewise, in his *Essays* Hunter states, "In Natural History we are often made acquainted with the facts, yet, do not know the cause. Therefore we are obliged to have recourse to experiment to ascertain the causes which connect with the facts, one leading into the other, making a perfect whole; for, without the knowledge of the causes and effects conjointly, our knowledge is imperfect."

Remember Hunter's eaten-away stomach specimen from chapter 2. That specimen now speaks volumes about Hunter's overall philosophy. First he ran innumerable experiments on digestion, carefully extending the experiments across many taxonomic classifications and recording significant correlations: feeding lizards, birds, and mammals foods tied to a string so that they could be pulled out of the stomach later; killing and dissecting animals to reveal the digestive process; measuring subtle changes in body temperature; analyzing the stomach muscles and gastric juices; and so on. Then he looked for general correlations and developed hypotheses that he tested. Contrary to his contemporaries, for example, he began to see that animal heat was not boiling or dissolving the ingested food. Rather, animal heat (the thermoregulation system) was powering the other physiological functions important to digestion, such as circulation and respiration. These kinds of inductive experiments also led him to the general principle that living tissues resist the actions of gastric juices. This general principle of life, this causal matrix, could be tested with new experimentation, such as detailed comparisons of circulation and respiration in cold- and warm-blooded animals and detailed postmortem examinations of the interaction between gastric fluids and stomach tissue.

Hunter would mercilessly prod his friend and former student Edward Jenner, for example, to make important observations of hedgehogs. Jenner, who lived in Gloucestershire, was frequently cajoled by letter to study specific habits of the hedgehog. But this was not the observe-and-classify method of the old-school natural history. Hunter sent Jenner a thermometer that he constructed for taking hedgehog temperatures during periods of hibernation. This was a more invasive, experimental natural history. As Hunter learned more about the natural history of hedgehogs, he learned more (because of his new physiological mind-set) about the principles of life (the phenomenon of animal heat). And the more he learned about these principles, the more he understood the specific animals in question. All this was related to Hunter's medical studies as well, because, as an early diagnostic surgeon, he called for medical professionals to move beyond the symptoms (the outward appearances) of maladies toward the inner causes of those symptoms.

Even the ultimate resister of explanation and system, the monster, was organized in Hunter's mind (and museum) according to principles (fig. 4.8). Cabinets were assigned for monsters due to "Abnormal Situation of Parts," for those resulting from "Addition of Parts," for those resulting from "Deficiency of Parts," and so on. John Hunter could not yet say what caused some of these particular categories of malfunction (there were deeper genetic and developmental engine rooms that would not be discovered until the twentieth century), but he never threw up his hands in frustration, declaring it to be the handiwork of a supernatural force.

This brings us back, finally, to the bizarre case of Thomas Lane and his fetus "tumor." While Hunter himself died some twenty years before this provocative specimen made its way into his collection, it was truly the appropriate place for Nathaniel Highmore's donation. It would be a very long time before anybody really understood such a strange phenomenon (indeed, it is not perfectly understood today), but Hunter's revolutionary approach to nature (metaphysically materialist, and methodologically causal and experimental) paved the way for science to bring even the freakish under its rational umbrella. Simply put, the female monster-fetus found inside Thomas Lane's gut was his sister.

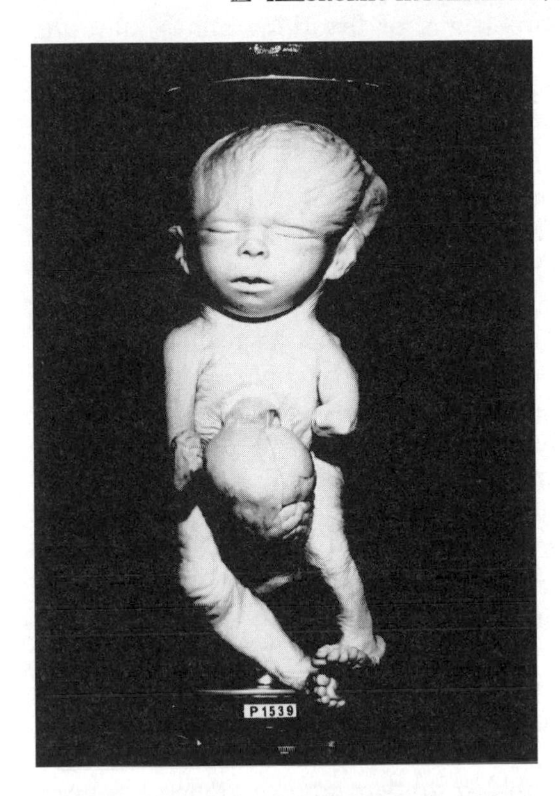

Figure 4.8. A teratological "monster" fetus, from Hunter's collection. (BY PERMISSION OF THE ROYAL COLLEGE OF SURGEONS, ENGLAND)

Neither Hunter nor Highmore nor anybody else from those days could have known it (since the mammalian egg wasn't even discovered until 1827), but Thomas and his sister started out as monozygotic twins. It is a rare occurrence that independent siblings fuse together and result in conjoined (Siamese) twins (one in forty-two thousand deliveries). Inside the single egg's blastodermic vesicle, the inner cell mass separates imperfectly into two conjoined parts. The precise cause of this cell mass separation is unclear, but it is believed that subtle environmental changes (such as changes in temperature or oxygen level) can result in twinning. Conjoined twins appear to be incomplete separations of an ordinary monovular twinning process. It is rarer still that such abnormal fusion should produce unequal conjoined twins, where one of the twins' development is stunted in utero and becomes a much smaller "parasite" on the larger, more developed, "autosite" fetus. While rare, these cases of unequal conjoined twins have been well documented throughout history.

A Genoan man, for example, named Lazarus-Joannes Baptista Colloredo, born in 1617, displayed himself throughout Europe. From his torso hung a shrunken upside-down twin with ill-formed arms but a well-formed head. The parasite twin showed some rudimentary signs of autonomy—signs of breathing and a salivating mouth. But according to reports, the parasite never opened its eyes, nor did it eat anything. The book *Medical Curiosities* reports of an "Italian child of eight who had a small head proceeding from under the cartilage of the third left rib. Sensibility was common, pinching the ear of the parasitic head causing the child with the perfect head to cry. Each of the two heads received baptism, one being named John and the other Matthew."

But even stranger still are the cases like those of Thomas Lane, wherein the parasite sibling (usually ill formed) actually lives inside the autosite. For unclear reasons, probably connected to vascular supply, an embryo that begins as an equal twin can become stunted and then internalized by their sibling.

What is particularly peculiar about the Thomas Lane case is the fact that he and his internal sibling were of different gender. This is unheard of with monozygotic twins, and it suggests two possibilities. Sexual differentiation is not simply a matter of chromosomal makeup; a fetus can be genetically male (XY), but if the developing endocrine system does not produce enough male hormones, the masculine genitalia can fail to develop. The early embryo of humans is sexually neutral and is capable of going either way depending on the chemical triggers that it receives later in gestation. Notice here that the ambiguity of gender itself is a fundamental stumbling block for the classifying mind, which seeks to put things in either/or categories. Between the seventh and the tenth weeks the embryo's gonads begin to transform into ovaries or testes, and if the testes fail to secrete proper amounts of androgens (male-producing hormones), then female genitalia result. So one possibility is that Thomas's twin became arrested in its development before it received its hormonal triggers and thus represented a genetic male without male genitalia. The other, less likely possibility is that Thomas and his sibling were not monozygotic but fraternal twins, separate fertilizations of two different eggs. But perhaps due to a glitch in the blood supply system of his sister, Thomas absorbed her into his own developing body.

The mystery of the Thomas Lane specimen was solved not by Hunter himself nor by his immediate successors. But it was the spirit of Hunter's generation that led biology to pierce through the veil of appearances (even the rare "monsters") to the hidden causal mechanisms. Hunter, by design or by dint of his personal dilettantism, annexed the growing experimental physiology to purely descriptive natural history. Both the outside and, now, the inside of the animal were seen as sets of adaptational relations.

While Hunter's and Cuvier's theories and collections reached beyond the immediate observables of nature to the causes, they did not reach (like so many of their predecessors' and contemporaries') toward the Deity. On questions of origins, both men remained formally silent and felt quite keenly that everyone else should probably remain silent as well. Following the skeptical lead of John Ray, Hunter and Cuvier figured the purposes of the Deity to be beyond the realm of human knowledge. God might have created these adaptations (and both thinkers believed that probably he had done so), but, from the perspective of logic and sense data, it was also possible that a giant cosmic spider had spun the world from its anus, adaptations and all. And lest the reader think that I've imported this impious possibility into the late eighteenth century, I hasten to point out that this ancient Hindu idea, of a world-making arachnid, was offered up by David Hume (1711–1776) as an equally rational alternative to the "design argument." For thinkers such as Cuvier and Hunter (interestingly, Hume was related to Hunter by marriage), metaphysics in general was going off the agenda.

Cuvier talked of "final causes" and "purposes" in his writings, but these were rhetorical devices designed to direct us from a specific animal character to a specific survival function. A fanglike incisor, for example, leads the naturalist to infer a life of carnivorous predation. Exactly how character and function had gotten linked together was shrouded in mystery, but that they were linked together was crystal clear. And, most important, good science could be done, Hunter and Cuvier believed, without ever entering into this particular question of how. In his *Essays* Hunter says that "to attempt to trace any natural production to its origin, or its first production, is ridiculous; for it goes back to that period, if ever such existed, of which we can form no idea, viz. the beginning of time."

Refusing to place God explicitly at the helm of nature was not a denigration of religion. The territorial boundaries of religion and science were being redrawn during this period. Intellectuals such as Hunter, Cuvier, Hume, and Immanuel Kant (1724–1804), were distinguishing two autonomous realms: the study of nature and of human beings' relationship to it fell under the purview of science (natural philosophy), whereas the study of morality and of human beings' relationship to each other fell under the purview of religion. For these thinkers, one did not learn anything about God or morality by studying nature, and one did not learn anything about nature by invoking the spirit realm.

I wandered a bit more through Cuvier's collection and eventually found Heidi staring at a shockingly bright orange-and-red orangutan specimen. The glass case was part of the "Respiratory" series, and the orang was splayed out at the bottom of the case. Since it was only the respiratory system that was being exhibited, the animal's lower half had been removed. This method of truncating specimens to fit them into the limited exhibit spaces is very prevalent in smaller museums, such as those of Hunter and Cuvier. There is something unnerving about these amputated organisms, particularly here at Cuvier's gallery, where the dissecting prowess does not match Hunter's clean artistry. Something of the violence of the saws and blades is faintly suggested by the raggedy ends of specimens, a violence that is mercifully absent in Hunter's meticulous preps. Perhaps, in a more subtle and insidious way, Hunter's wet preps are creepier because of their seemingly inorganic, highly manicured quality.

Heidi and I left the museum, emerging out of the catacomblike darkness into the brilliant light and color of the gardens. Making our way on the wide gravel paths, we leisurely passed by lavish displays of dahlias and geraniums, a gorgeous rose garden, hundreds of varieties of irises, and then earthenware pots of medicinal herbs. We came across a jaw-droppingly massive cedar tree. A plaque explained that it is one of the oldest trees in Paris, planted by Bernard de Jussieu (the fellow who detected Linnaeus amongst his botany students) in 1734. Here was a living link to the days when Buffon walked these paths. Linnaeus, Buffon, Cuvier, Lamarck, Geoffroy—all of them orbited around this tree, literally and figuratively. The cedar, not just a symbol but an actual expression of nature itself, grew quietly, inexorably onward, while the human mind chronologically circled it for answers. We paused reverentially before this arboreal bridge to the past.

After eating at an outdoor café on the west side of the Jardin, we decided to do a museological exploration of the nearby Grande Galerie de l'évolution. As we approached the Grande Galerie I reflected on the taxonomy issues again. Naturalists such as Hunter and Cuvier steered biology away from the Linnean focus on form and toward a new focus on function. But eventually there were nagging problems with this functional way of carving up the categories. Many animal-body plans seemed obviously similar, but these animals did not have similar functions—they did not live in similar environments, they did not eat similar food, they did not behave similarly, and so on. Homologies, such as the jigsaw skulls of mammals and birds that we saw in chapter 1, made no sense from the strictly functional perspective (certainly mammals have to squeeze big heads through small openings during birth, but birds don't). Cuvier's colleague at the National Museum, Geoffroy, was increasingly impressed with these nonfunctional homologies of the animal kingdom and wanted to display animals accordingly. Geoffroy even went so far as to promote the idea that mollusks and humans have homologous visceral layouts; if you bend a human so that the back of the head is touching the buttocks (don't try this), the human's internal organs will resemble those of a cuttlefish. After 1830 biology was all over the map. It was successful locally, in its various subbranches, but there was no global unity to the proliferating research programs.

With the publication of Darwin's *Origin of Species* in 1859, everything changed. Homologies that seemed to run counter to adaptation now made sense because distant common ancestors had left their marks on their currently very different descendants. And the adaptations that Hunter and Cuvier had focused on were now tied to the process of natural selection (the ultimate engine room). Moreover, the natural classification system itself (still primarily Linnean), with its nested differentia, was now simply a by-product of the evolutionary tree, mere frosting on the temporal cake. It was a revolution that museums would have to both follow and lead. Exhibition practice would have to follow the new theoretical principles, but, as public educational institutions, museums would have to lead the popular imagination to grasp this "new" nature.

CHAPTER 5

EXHIBITING EVOLUTION: DIVERSITY, ORDER, AND THE CONSTRUCTION OF NATURE

SPENDING TIME PERUSING AND "DISSECTING" museums from the late eighteenth and early nineteenth centuries gives one some refreshing distance from the more familiar evolutionary exhibits. In the late nineteenth and early twentieth centuries American and European museums primarily concerned themselves with educating the public about the theory of evolution, explaining the principles and displaying the evidence. Currently, a significant amount of museum space is still given over to the pedagogy of evolution. We are so accustomed to evolutionary displays that we often fail to appreciate their underlying logic. Stepping back and examining one's own paradigm is, after all, almost as hard as tasting one's own tongue. As we saw in chapters 3 and 4, organizing and displaying specimens is a way of visualizing the invisible, of making ideas palpable. There are rich invisible cultures that, if analyzed properly, emerge out of the material culture of contemporary evolution exhibits.

My first contemporary exhibit "dissection" was performed on the beautiful, and recently renovated, Grande Galerie de l'évolution of Paris's Muséum national d'histoire naturelle. I eventually flew back to London for a comparison with the prestigious Natural History Museum's handling of evolution, then on to the American Museum of Natural History in New York, and finally back home to Chicago's Field Museum and their "Life over Time" exhibit. The best way to proceed now, however, is not chronologically but thematically. Different countries curate evolution differently, and these variations, sometimes striking, sometimes subtle, reveal important invisibles.

In this chapter I want to use these four museums in three different countries as a medley of case studies. Studying these different cases reveals some hidden cultures of collecting and representing and allows me to explore the differences of rhetorical emphasis (i.e., which version of nature—diverse or orderly—is being presented) and some interesting "nationalist" tendencies in museum display (i.e., how each museum is a reflection of culture and country). In the following chapter, I continue my search for hidden cultures by focusing on the Field Museum and a Chicago-style lesson about God and gambling. But before I tackle the two themes above, however, it is necessary to review some relevant principles of Darwinian evolution. How does it work, and why is it good science?

Lamentably, evolutionary science has become a significant blind spot for most American students, and even the seemingly initiated ones are remarkably uninformed on the specific mechanisms of evolution. The idea that the schools are overrun with dangerous Darwinians, peddling their unholy heresies, is egregious hyperbole. Consider some of these depressing facts about the state of biology education in America. In a study conducted in the late 1980s by the International Association for the Evaluation of Educational Achievement, seventeen-year-olds from almost two dozen countries were comparatively assessed. American teenagers ranked in the bottom 25 percent in biology. In a three-state survey of undergraduates (described in the *Chronicle of Higher Education*, November 1986), more than half the students polled claimed that they were creationists (and a third also claimed to believe in ghosts, communication with the dead, aliens, Bigfoot, etc.). In some revealing Ohio surveys, it was learned that only 12 percent of high-school biology teachers could correctly identify a summary statement of evolution theory from five multiple-choice options. And in a similar survey, only 2 percent of school-board presidents were able to properly identify evolution theory at all. In a 1990 Gallup poll of 1,236 American adults, 65 percent believed in Noah's ark and the biblical flood, and 41 percent believed that dinosaurs and humans lived at the same time. In 1997 the Illinois state board of education refused to include any reference to evolution in its new statewide standards, and in 1999 Kansas eliminated the term *evolution* from its school science standards. All this makes me think back to Dr. Albert Eide Parr (chapter 1), who, when addressing the fiftieth-anniversary audience of the Field Museum in 1943, claimed that the museum's earlier goal, of educating the public to the truth of evolution, had already been achieved. He seems to have been premature in that assessment.

Darwin, contrary to common confusion, did not discover evolution. The general idea of evolution (a term that originally meant the "unrolling" of preformed creatures, like the opening of nested Chinese boxes) came to mean the transformation of populations and species through time. The idea that populations of organisms change over time was *not* a Darwinian breakthrough; isolating the processes that actually produce evolution was Darwin's true contribution. Beginning from the observation that all offspring vary slightly from their parents and their siblings, he asks (in chapter 4 of the *Origin of Species*):

> If such [variations] do occur, can we doubt (remembering that many more individuals are born than can possibly survive) that individuals having any advantage, however slight, over others, would have the best chance of surviving and of procreating their kind? On the other hand, we may feel sure that any variation in the least degree injurious would be rigidly destroyed. This preservation of favourable individual differences and variations, and the destruction of those which are injurious, I have called Natural Selection, or Survival of the Fittest.

Darwin made three relatively uncontroversial observations and, combining them, turned them into the most controversial science of the last two centuries. The first observation is that organisms create many more offspring (eggs, seeds, live young, whatever) than actually make it to adulthood. If just one season's worth of deposited mosquito eggs actually made it to adulthood, we'd all be swimming in an earth-blanketing sea of bloodsuckers. Or ask Australians, where the whole ecology is still being devastated because a few rabbits were introduced in 1859, if there is any truth to the old saw "multiply like rabbits," and the answer will be an emphatic assent. One of the principal strategies of nature seems to involve spilling your seed wherever, whenever, and as often as you can, so that the law of averages will grant a few successes. Very rarely, as in the case of rabbits in Australia, an organism finds itself in a new niche where the usual population checks do not exist. And when that happens, you can witness Darwin's fecundity principle plain as day. Darwin even takes a relatively slow reproducer such as the elephant and shows how utterly excessive all reproduction rates are. It is safe to assume, he argues, that if the elephant "begins breeding when thirty years old, and goes on breeding till ninety years old, bringing forth six young in the interval, and surviving till one hundred years old; if this be so, after a period of from 740 to 750 years there would be nearly nineteen million elephants alive, descended from the first pair." Natural checks, including scarcity of resources and predation, ensure that the multitudes at the starting gate of life become severely whittled down before the finish line of adulthood.

The next simple observation is that all organisms vary naturally.

No two individuals of a species are exactly the same (except in the rare case of monozygotic twins). Offspring certainly resemble their parents more than they resemble nonrelatives, but they also deviate in myriad ways: their skin color, eye color, blood type, size, shape, chemical composition, predisposition to diseases, behavior, and so on all vary. Nothing very shocking or controversial here.

The last piece of this elegantly simple puzzle is this: Any genetic variation that happens to be advantageous would tend to help the organism reach sexual maturity and thus be passed down to the following generation. *Advantageous* is a completely relative term. The well-known Spencerian parlance "survival of the fittest" still misleads people into thinking that Darwinism is just an aggressive arms race in which bigger muscles and sharper teeth and claws inevitably win out. But fitness merely means the ability to leave more offspring. Yes, some animals really do struggle with each other for existence, but *struggle* is just a term of metaphorical convenience. Darwin says, "Two canine animals, in a time of dearth, may be truly said to struggle with each other which shall get food and live. But a plant on the edge of a desert is [also] said to struggle for life against the drought." And it can also be said that fruit-bearing plants compete with each other to attract birds to eat and disseminate their seeds. But all that is meant by this is that plants that vary in ways that bring more birds to their fruit (perhaps by smell, or coloration, or meatiness of fruit, or increased numbers, or enlargement, or diminishment, or whatever) are better able to pass on their genes.

These three points summarize Darwin's mechanism of chance variation and natural selection. This is the engine of biological evolution; it is what makes organisms adapted to their conditions of existence. It is important to add here that Darwin himself did not know anything about genetics, because *nobody* knew there were such things as genes until later. It was not until the 1880s, for example, that biologists even felt confident in claiming that something called a chromosome had a role in heredity.

The obscure botany experiments of an Austrian monk named Gregor Mendel (1822–1884) were rediscovered in 1900, and this led to a greater understanding of the logic of inheritance and a new wave of biologists called geneticists. The geneticists believed in a theoretical particle or substance called a gene that existed at the subchromosomal level and made sense out of Mendel's predictive patterns of heredity and variation. Darwin's claim that chance variation delivered up the raw material for natural selection to work on was refined and elaborated as early-twentieth-century biologists discovered that the variations were due to mutations and recombinations of the genes. Then came the discovery of the DNA double helix by James Watson and Francis Crick in 1953. As Julian Huxley (1887–1975) argued in his 1963 edition of *Evolution: The Modern Synthesis*, the discovery of DNA reiterates the Darwinian principles at a smaller scale and sheds greater light on the macro changes. He points out that the "existence of an elaborate self-reproducing code of genetical information ensures continuity and specificity," which means that DNA explains how homologies get passed along and why dogs give birth to dogs, humans give birth to humans, and so on. Having a certain code means having a certain kind of body, and this means having a certain place in the taxonomic table. Huxley goes on to show how DNA makes evolution inevitable by stating that "the intrinsic capacity for mutation [at the base levels of adenine, thymine, cytosine, and guanine] provides variability; the capacity for self-reproduction ensures potentially geometric increase and therefore a struggle for existence; the existence of genetic variability ensures differential survival of variants and therefore natural selection; and this results in evolutionary transformation."

The significance of all these amazing advancements in biological knowledge is that while they are very important in their own right, they are mostly just a filling-in of the empirical gaps that Darwinian evolution had already recognized. The conceptual framework that Darwin laid out continues to provide the foundation supporting all of twentieth-century biology. We know much more now about some of the nonadaptive mechanisms of evolution (mechanisms besides natural selection, such as the founder principle, genetic drift, and generic structural trajectories), but all of these are just enrichments of Darwin's theory of descent with modification.

One might well press this account of evolution as changes in gene frequencies by asking how such a process creates new species. How does one interbreeding population get transformed into two different and reproductively isolated populations? The most prevalent manner by which genetic variations are shaped into resulting new species is geographical isolation. If a single population of plants or animals is fractured into two groups by some geological process or environmental event and the groups cannot rejoin each other for many generations, then their genetic variations may accumulate to such a degree that an interbreeding "reunion" of the groups becomes impossible (remember Alfred Wallace's argument in chapter 1 that the deep waterway between Bali and Lombok effectively isolated the evolutionary pathways of Australia and Southeast Asia). When small animals or seeds get blown freakish distances by the wind or washed up in new environments; when floods, quakes, and fires separate populations; or when the normal geological processes of river formation, mountain building, erosion, and plate spreading and subtending block one population's access to another, then the constantly accumulating mutations of the estranged populations can make them reproductively alien to each other. This alienation of two populations can build up in subtle genetic ways, wherein the fertilized egg naturally aborts or fertilization never occurs, or it can be in macro mechanical ways, such as with the mismatched genitals of Chihuahuas and Great Danes (an example of human-engineered isolation). This is how Darwinian evolution works. But *why* is evolution theory good science?

The reason why scientists have embraced Darwinism is because it explains the apparent order in nature, and it sheds light on so many puzzles. It is worth noting a few of these conundrums, and also examining the successful manner in which natural selection explains the age-old, contentious puzzle of human skin color.

For example, remember that Cuvier's principal law of nature was that animal structures always fit their particular functions (their activities in a specific environment). This lawlike correlation allowed him to make deductions from anatomy to physiology and vice versa. While Cuvier refused to draw any explicit theological implications from his law, many creation-minded naturalists inferred God as the cause of this correlation and then, of necessity, were logically forced to sharpen plain old adaptation into "perfect" adaptation. After all, the thinking went, a perfect God would not create an imperfectly adapted creature.

Darwin noticed many embarrassing exceptions to this doctrine of perfect adaptation. In chapter 6 of the *Origin of Species* he reminds readers of curious cases where structures and functions do not align. Webbed feet of ducks and geese, for example, seem to be obvious adaptations for swimming. Yet there are species of geese that never go near the water, despite their fully webbed equipment, and, contrariwise, there are some aquatic birds that do not possess webbing at all. Darwin says that "he who believes in separate and innumerable acts of creation may say, that in these cases it has pleased the Creator to cause a being of one type to take the place of one belonging to another type; but this seems to me only re-stating the fact in dignified language." In other words, this creationist explanation is not an explanation at all. The uniform response "God wants it that way" is a rhetorical foil masquerading as an explanation. Unless the naturalist can suddenly do divine psychology, the solution to a natural puzzle quickly retreats into the inscrutable supernatural domain. But, Darwin explains, someone "who believes in the struggle for existence and in the principle of natural selection will acknowledge that every organic being is constantly endeavoring" to fill any available niche that it can. So if an animal can flourish by moving into a new lifestyle (a habitat that is different from the one that originally shaped its structural equipment), then it will, and the naturalist out in the field will consequently witness a slight slippage or incongruity between structure and function. The slippage makes sense only from the perspective of Darwinian principles. But this case is a mere trifle compared to the other explanatory accomplishments of evolution theory.

Here is a short inventory of Darwinian success stories, features of nature that become illuminated by evolution-based explanation but remain dumb mysteries without it. First, evolution makes excellent sense of the fossil record. Stratigraphic sequences of older and younger rocks were established before Darwin, but evolution explains why we find a succession of primitive life forms in older or lower rocks and advanced life forms in younger or higher rock strata. As time passes and environments change, some species go extinct, while other species originate and flourish, and all this is evidenced in the fossil sequence. This point should not be confused with a Whiggish (progressivist) view of evolution. The terms *primitive* and *advanced* should not be understood as value judgments; organisms are not better because they appear in younger rock strata. An advanced organism is one whose specialized anatomy presupposes certain common structures as precursors. *Primitive* and *advanced* are relative terms only, and they are totally dependent on context (e.g., the trait of a synapsid opening in the skull is primitive for the Mammalia taxon, but the same trait is advanced for a more general taxon such as Vertebrata).

Natural selection also explains the adaptive fit of organism and environment. Without appeal to supernatural design or Lamarckian inheritance of acquired traits, natural selection can explain how organisms fit their geological environment and how certain species come to be adapted to other species. For example, plants that are preyed upon by herbivores can evolve increasingly toxic chemistry as a quid pro quo, and the digestive systems of the herbivorous insects can, over many generations, evolve toleration for these toxins. Passionflower vines produce toxins that ward off most herbivorous insects, but the butterfly *Heliconius* has adapted to the diet by developing digestive enzymes that break down the toxic compounds. Predation tactics and defensive counteradaptations can adjust the anatomy, chemistry, and behavior of one species to another. It can also lead to mimicry, in which a predator or prey species may gain advantage by developing a superficial resemblance to another species (e.g., when disturbed, the hawkmoth larva can puff up its head to gain a striking resemblance to a poisonous snake; it will even wave its head back and forth, hissing like a snake). Symbiotic transformations, as well as the more antagonistic ones, are governed by this same process of mutation and selection. All these adaptations are straightforward consequences of Darwinian evolution.

Next, the amazing homologies (previously mentioned in chapter 1) that we find across taxa are rendered more intelligible in light of evolution theory: the partitioned jigsaw skulls of mammals and birds, the common forelimb structures of birds, lizards, whales, cats, horses, and humans. These deep-structure similarities that underlie very different kinds of animals reflect a distant common ancestor in the same way that the English word *pensive*, the French word *penser*, the Italian word *pensare*, and the Spanish word *pensar* suggest a common ancestor (the Latin word *pensare*).

Next, scientists recognize that evolution theory makes sense of the varying degrees of genetic similarity between animals. In addition to the well-known cases of molecular fraternity between blood proteins in humans, chimps, and gorillas (for example, out of the 146 amino acids that make up the beta chain in hemoglobin A in both humans and gorillas, there is only one amino acid that is different between the two species), the very existence of a common and ubiquitous building block of all life, called DNA, is itself a chemical corroboration of evolution.

Next, the existence of seemingly useless organs and structures is unintelligible without evolution, but quite coherent with it. Wisdom teeth and the appendix are well-worn examples of structures that served a function in prehuman history, but some lesser-known cases include the useless femur and pelvis in whales (indicating their terrestrial quadruped ancestors), a similarly useless pelvis and femur structure hidden within the bodies of primitive snakes such as boa constrictors (indicating that snakes evolved from four-footed lizards), insect wings that are utterly incapable of flight, useless horse's toes, and other such obsolete structures. These once adaptive organs and structures are simply hangers-on due to genetic inertia; the old functions are gone, but the structures are not forgotten (genetically speaking). In chapter 13 of the *Origin of Species*, Darwin again mentions (with some frustrated humor) the nonexplanatory character of competing theories. "In works on natural history rudimentary organs are generally said to have been created 'for the sake of symmetry,' or in order 'to complete the scheme of nature'; but this seems to me no explanation, merely a restatement of the fact." He goes on to lampoon this way of thinking with a reductio ad absurdum: "Would it be thought sufficient to say that because planets revolve in elliptic courses round the sun, satellites follow the same course round the planets, for the sake of symmetry, and to complete the scheme of nature?" After our survey of naturalism in chapters 3 and 4, we can appreciate Darwin's comments here. He is fulfilling the causality paradigm of natural history (begun by Hunter and Cuvier), and his expedition into the engine room is a move beyond the anthropomorphic, aesthetic approach to nature.

In addition to the superiority of evolution theory for explaining useless (now vestigial) organs, we have to recognize the new light that the theory throws upon the "intermediary" animals so celebrated by previous naturalists such as Buffon. The slime hag (a lower chordate) that we mentioned in chapter 4 is a kind of intermediary between invertebrates and vertebrates, a species with one foot in each classificatory category. And of course, in addition to living intermediaries, we have fossil intermediaries such as the archaeopteryx (part bird, part reptile), which we discussed in chapter 1. There are many other phenomena (and even entire scientific branches such as pharmacology, agriculture, ecology, etc.) that become more intelligible in the light of Darwinian evolution, but this short list is enough to demonstrate that scientists have not embraced evolution in a flippant or conspiratorial fashion.

While the rhetorical harangues are rather tedious, there is a positive by-product of the creationism/evolution debate. Theoretical and legal battles over what subject matter should be taught in biology classrooms have resurrected interesting questions about what distinguishes science from nonscience (an important issue known to philosophers of science as the demarcation problem). With evangelical Christians and unctuous politicians claiming that evolution theory is no more scientific than creationism, scientists and philosophers have had to articulate more explicitly the criteria by which something actually qualifies as a science. This is all for the good, though at first the complexity of the criteria tends to confuse people who were raised on the overly simplified version of science as an infallible truth machine (that is, most of us).

Briefly, there are several tests for determining whether a candidate theory or idea is scientific. It's not written somewhere in stone that a theory must pass *all* of these tests, but it seems that some combination of these criteria is always found in the activities we call scientific, and the criteria generally are lacking in other activities such as art, religion, and so on.

A scientific theory must be capable of failing, that is, of being falsified by experiment. If a person frames a prediction or claim in such a way that it cannot fail, then, strictly speaking, it cannot be tested. Evolution theory is a falsifiable or testable theory because it commits to many empirically verifiable claims. Some sixty million years passed between the extinction of dinosaurs and the rise of hominids. If hominid fossils began to be discovered in the same geological strata as dinosaur fossils, thus suggesting that they were around at the same time, this would probably throw very serious doubts on evolution as we know it. Likewise, if it could be shown that injurious mutations could be spread throughout populations indefinitely (a kind of orthogenesis), then scientists would have to rethink the validity of natural selection.

In reality, the relationship between evolution theory and the falsifiability criterion is slightly more complicated than stated above. Specific phylogenetic patterns are falsifiable, but macroevolution, as a general process, may not be so falsifiable. This is because the explanatory theory of macroevolution has become more of a paradigm (an embedded explanatory presupposition that functions as an organizational framework) rather than a specific hypothesis that stands or falls according to specific experimental outcomes. There is nothing bogus or deceptive in this paradigmatic status of evolution. Like physics, chemistry, geology, and other disciplines, the practice of science is more complicated than the strict hypothetico-deductive model that we all learned in school. Sometimes, especially when an explanatory framework has been very successful in the past, scientists will look for ways to reinterpret potentially falsifying data (negative findings or experimental outcomes) rather than jettison the explanatory framework itself. This is not a disingenuous practice; it is a reasonable response to data anomalies. If enough of these anomalies eventually accrue and cannot be integrated into the parameters of the paradigmatic assumptions, then those metatheoretical presuppositions begin to undergo reexamination and possible revision—perhaps even revolution. Suffice it to say, however, that the most interesting evolutionary claims are local and testable rather than global and philosophical.

A legitimate scientific theory should also be fruitful. A good theory makes new predictions; it anticipates, pursues, and corroborates future facts. More than just accounting for previously encountered facts of nature, science makes bold claims about previously unknown phenomena and then goes looking for them. Evolution theory produces fruitful predictions. It predicted that the chromosomes and proteins of related species should be similar; indeed, the very concept of descent with modification presaged the discovery of some unit of heredity (the gene). Also, while many transitional fossils have not been and probably never will be found (due to the nature of fossilization and allopatry), it is true that a substantial number of "missing links" have been predicted and subsequently discovered.

Besides being a fruitful research program, evolution's explanatory successes in areas such as paleontology, microbiology and genetics, adaptationism, homology analysis, taxonomy, and suchlike illustrate the amazing scope of the theory. And evolution is ontologically simple. That is to say, no mysterious forces or supernatural entities have to be assumed or proffered in order for its explanatory power to work; natural replicators, together with limited resources, constitute a relatively seamless system. Last, Darwinism is conservative in the sense that it harmonizes well with what we know from other fields, such as contemporary chemistry, physics, cosmology, medicine, geology, and other sciences.

Briefly sketching a specific example—skin coloration—reminds us of the explanatory power of Darwin's theory. It also raises interesting questions about classification theory and practice. These points will be helpful as we turn to the contemporary museum displays.

It seems that human beings have always recognized racial categories, and we have often used racial traits as classificatory boundaries. Sometimes this is a relatively harmless demarcation, though sometimes it is motivated by a nasty political agenda. Skin coloration, for human beings, has traditionally provided the most obvious signifier of group membership. The visual discrimination of human physical differences (a category discrimination, not a value discrimination) seems almost physiologically automatic. However, we are now learning, mostly through evolutionary science, that our folk categories of race, based on skin color, are misleading. Evolution forces us to do away with the old notion of essential, changeless categories and replace it with the idea of varying populations. Races are fuzzy sets, not traditional either/or categories, and skin coloration is not the obvious signifier that our folk brains want to contend that it is. Australian Aborigines, for example, are frequently believed to be closely related to Africans because of the similarity in skin coloration. But they are actually very distantly related groups, genetically speaking. Aborigines are genetically much closer to Asian peoples than to Africans, and this is completely understandable (though perhaps frustrating to folk-brain classifications) when we realize that their ancestors moved from their site of origin in Africa into the Middle and Far East, then down through Southeast Asia, and finally into Australia.

Granting that racial variations such as skin coloration are well-recognized phenomena (though misleading for taxonomy), how does evolution explain them? Darker skin is caused by greater production of the pigment melanin by melanocytes in the basal epidermis. Granules of the melanin produced by these special cells migrate into epidermal cells and move up to cover the nucleus, protecting it from the harmful effects of ultraviolet light. If you look at the distribution of strongly melanized (dark-skinned) people around the world, you find that they tend to be clustered near the equator. Generally speaking, the closer a population is to the equator, the darker their skin is. This correlation of coloration and environment is no accident, nor is it the product of a designing Deity. It is the product of natural selection on a population that must contend with the locally intense threat of skin cancer.

When solar radiation is quite intense, a light-skinned person burns. Repeated sunburning of a person with few melanocytes can lead to degeneration of the dermis and epidermis, and cancer can ensue. Someone who has darker skin pigmentation is better protected from the cancer-causing forms of solar radiation. This fact is empirically borne out by the fact that death rates from skin cancer in whites are three times higher than in blacks. And cases of skin cancer are even higher for whites living in the southern United States, slightly closer to the intense solar radiation found in the equatorial region, than whites in the north. It is generally believed now that in equatorial populations, genetic variations producing dark skin were selected for (darker people lived long enough to reproduce and pass on their traits), and genetically lighter-skinned folks were selected against. Computer models of the selection process estimate that it would take between twenty-four thousand and forty-five thousand years for a hypothetical population to go from white to black, but of course this is only a blink compared with the millions of years of human evolution.

Why, given the protective character of melanization, would originally dark-skinned populations evolve lighter skin? Here again we have to think about the environment and the possible usefulness of light skin. In keeping with the previously mentioned distribution of skin color around the globe, we find that lighter-skinned populations are generally more prevalent the further you get from the equator. People whose genetic variations produced lighter skin colors were selected for in less sunny environments, while people with darker skin were selected against.

In order to understand this fact, one has to appreciate some specific features of the sun. Radiant energy spreads out from the nuclear conflagration at the speed of light, making up a spectrum of wave forms that bombard the earth's atmosphere. This spectrum includes

X-rays, ultraviolet rays, visible light, infrared rays, and radio waves. The ozone layer blocks many harmful radiation wavelengths, but the narrow wavelength band between 290 and 310 nm (the same band width that can cause cataracts, sunburn, and skin cancer) is necessary for the human body to synthesize vitamin D. Unless one is willing and able to eat 23 g of sardines or 100 g of herring a day (to name two other sources of vitamin D), then one is probably going to have to get one's vitamin D directly from the sun's ultraviolet rays.

Sunlight falls on the skin and converts 7-dehydrocholesterol into cholecalciferol (vitamin D); this chemical is then transported from the skin to the liver, where it is hydrolized, and after further hydrolization in the kidneys it enters blood circulation to be used during calcium metabolism. Simply put, sunlight makes good bones. Without the proper amount of sunlight, human beings do not develop healthy bones, and in fact there is copious evidence relating higher incidences of rickets to populations living in regions with short summers and long winters.

Can this environmental pressure have an evolutionary effect? Most definitely. Without enough vitamin D synthesis, one contracts rickets; if a woman is stricken with rickets, then her pelvis will bow inward; if her pelvis becomes deformed, then the smaller passageway will not allow for proper childbirth, and both mother and baby will perish. Thus a direct selection process works to eliminate those members of a population who do not synthesize sufficient levels of vitamin D.

Dark skin blocks some 95 percent of the UV radiation from reaching the deep skin layers where vitamin D gets made, but if you're in the sun all the time (relatively speaking), then you still get enough vitamin synthesis. If your skin is dark (a natural selection result of the sunny environment) but your population migrates away from that original environment to a region with significantly less sunny skies, then you've got trouble. In a low-insolation environment, the selective pressure of skin cancer dissipates, only to be replaced by an equally deadly threat of radical bone disease. Too much sun is bad; too little sun is bad. Consequently, genetic variations producing lighter skin will be selected for in the temperate climates, and variations of darker skin will be selected for in tropical climates. This is essentially the way in which natural selection and chance variation combine to create human skin colors. It is a well-researched and well-tested case of evolution in action.

Skin color is one of those human attributes that museums have traditionally used as the basis for exhibits. At the Musée de l'homme, directly across from the Eiffel Tower, one is struck by the nontraditional anthropology exhibits of human diversity. When entering the human diversity exhibit, you ascend a long stairway. At the top of this stair is a life-sized photograph of twenty different people of all ages and races, smiling and waving to you. There is nothing obviously interesting about this, except that they're all naked. And I don't mean cleverly naked, with well-positioned fig leaves or strategic airbrushing; I mean full-on naked. One would be hard pressed to find anything as frank and refreshingly honest in the puritanical United States. This photo and an adjoining one that focuses on skin color are labeled "Les Hommes different par leurs caractères sexuels" and "Les Hommes different par la couleur de leur peau," respectively. They simultaneously celebrate the unique differences of human people (not "specimens" in the sense of race-category representatives—these were definitely individuals) while also celebrating the egalitarian unity of being human. There is something very leveling about nakedness.

In addition to this nonhierarchical display of human kinds, there is a funny and ironic sequence of human skulls. The skulls are laid out sequentially so that the first one dates back almost a hundred thousand years. This controversial "Man of Qafeh" (discovered recently by French scientists in lower Galilee) places anatomically modern humans earlier than previously thought. The skull is displayed along with stones chipped for cutting purposes. The next skull is that of a forty-thousand-year-old Cro-Magnon and is accompanied by flint arrows. This is followed by the seven-thousand-year-old L'Homme de Villeneuve–La Guyard, accompanied by crude jewelry and a hammer. The next skull appears to be the final specimen in the consecutive series. It is the skull of French philosopher René Descartes, sitting majestically next to his *Collected Works*. This familiar Eurocentric sequence, here with Descartes at the zenith, visualizes an unambiguous invisible: The march of progressive evolution culminates in a Frenchman. But one additional specimen has been added, one that effectively democratizes the formerly hierarchical display. Next to Descartes's skull is a case where a video monitor sits amidst other modern accoutrements, including a remote-control device, a computer chip, and a credit card. On the screen of the monitor is the viewer's own face, filmed by a hidden camera. The visitor can feel like the zenith of evolution. All visitors who look into this exhibit, no matter their race, age, or gender, find themselves not only superseding Descartes, but reflected inside the very process of a developing nature.

Of course, there is no real acme to evolution. Evolution is not progressive, because the idea of "success" or "better" is completely relative to environment rather than absolute. The skin color analysis we just worked through is a case in point. But what a wonderful exhibit. Evolution is not something that happens to other people or other organisms; it is not occurring outside you or beyond you as a remote natural process. The video self-portrait specimen conveys the idea that you, and everybody else, are an equal part of the evolutionary process.

By rearranging specimens in this way (and selectively adding and subtracting them), curators can make specimens say different things to patrons. Sometimes, as in the Musée de l'homme, the new arrangements even self-consciously mock the old-style arrangements. In museums, nature is constructed rather than discovered. This is not to suggest total relativism, in which curators simply impose arbitrary hegemonies as they see fit. Museology, while being inherently political, is not utterly divorced from factual foundations. The nature that is constructed for public consumption (in natural history museums, anyway) is usually an excerpted and accentuated paragraph of the otherwise boundless book. When curators construct nature, they are certainly influenced by their respective paradigms, but this doesn't mean that they're making it all up. They are simply choosing and editing the empirically investigated bits of nature into a manageable story. By selectively emphasizing certain aspects of nature, the act of curating becomes both objective and subjective. When we examine arranging in this way, we taste our own tongues a little bit. Not only do we see nature, we also see ourselves seeing nature.

Looking at the history of the Hunterian Museum, for example, allows us to catch some of this constructing activity (and this activity, together with its conceptual blueprint, is the real back stage to any museum's front stage). As we saw in chapter 4, Hunter himself arranged his specimens to convey broad categories of similar animal functions such as reproduction or locomotion. The functions were abstracted out of their specific animal sites and rendered obvious as physiological frames of nature. For Hunter, the functionalist construction of nature radiated (without clear divisions) through the fields of natural history, physiology, pathology, and anatomy. But when Richard Owen took over the collection (as assistant conservator in 1827 and conservator in 1842), things changed. Same collection, different nature.

Owen was much more interested in the anatomical specimens (both current and fossil) than in the medical and physiological features of Hunter's collection. He became a prime mover in the burgeoning field of comparative anatomy, studying early on with Cuvier but eventually turning away from the Frenchman's exclusive emphasis on function. Some of the trustees of the Royal College felt that Owen's stress on anatomy and natural history, rather than medical topics, was taking the collection in the wrong direction. Actually, Owen's increasing taste for paleontology, along with the fracturing of the sciences during this period, eventually led him away from the Hunterian collection and toward the founding in 1881 of the Natural History Museum in South Kensington (an offshoot of the British Museum). But from the 1830s to 1856 Owen had major control of Hunter's collection. During this time Owen stressed, both in his museum arranging and in his public Hunterian Lectures, the morphological similarities of animal bodies.

Emphasizing anatomical homologies rather than functional similarities (analogies) was a break from Hunter's original plan. It was Owen who first celebrated the homologous braincases of birds and mammals. Hunter's original stress on unity through nature seems to have mutated in Owen's mind to mean unity of type rather than unity of function. The unity-of-type idea was a way of talking about the deep structural similarities that animals seemed to share (for example, the quadruped body plan underlies many different kinds of animal anatomies). But these deep homologies, which Owen called archetypes, were not seen as signs of evolutionary ancestors. Owen broke with his old friend Darwin and argued that these archetypes were somehow like transcendental Platonic Ideas rather than signs of common ancestry. Animals that were anatomically unified by a deep structure were seen as different permutations on God's original Idea. Just what degree of transcendentalism Owen subscribed to is a matter of debate, but what is clear is that Owen used Hunter's collection as a way of constructing a nature different from that of the original curator.

In 1945 debates about the scientific purpose of Hunter's collection were still raging. Shortly after the World War II bombing of the museum, curator G. Grey Turner became circumspect about the future function of the collection. Turner claimed that treating Hunter's specimens as a gateway for understanding taxonomy was an otiose agenda, as was a strictly natural history agenda. The new function of the collection, he argued, should be the understanding and improvement of human health (i.e., stressing the medical specimens, contra Owen). But while some of the trustees and Royal Society fellows of this period called for elimination of the nonhuman specimens, Turner rebuked them on the grounds that these exhibits also had a place under the new aegis of human health, saying that "there should be preserved precisely those skeletons that may one day help us understand the structure and function of man." Since no one really knows which of these will be useful in the future, Turner suggested, the trustees should save everything, but also rethink it all under a new purpose. The nonhuman wet preps too should be preserved, because "just fancy the interest of comparing the oesophagus of the giraffe with its insignificant counterpart in man! Perchance the secret of achalasia may be yet disclosed in the study of its neuromuscular mechanisms."

It is clear that collections are promiscuous bodies of evidence, available for the varied goals of diverse curators and eras. Different nations and generations use (and abuse) collections to highlight different theoretical objectives and different facets of nature. The meaningfulness and the usefulness of a collection is found in the relationships created between specimens, rather than the specimens themselves. Specimens are like words in the sense that their meaningfulness happens only within the context of rule-governed semantic systems. Charles Willson Peale articulated this insight in 1800 when he said, "[I]t is only the arrangement and management of a Repository of Subjects of Natural History, that can constitute its utility. For if it should be immensely rich in the numbers and value of articles, unless they are systematically arranged, and the proper modes of seeing and using them attended to, the advantage of such a store will be of little account to the public." Always keeping this constructive feature of museology before our mind's eye, let's now jump up to the present and examine some of the ways that British, French, and American museums arrange nature.

The Grande Galerie de l'évolution in Paris was opened originally in 1889 as the Galerie de zoologie. Like the Hunterian, it suffered major damage during World War II and was closed altogether in 1965. In the mid-1980s scientists and museologists began to envision a new institution rising phoenixlike from inside a restored earlier structure. The unifying theme chosen for the permanent exhibition was evolution itself. The state-of-the-art gallery made its debut in the summer of 1994.

From the time you enter until the time you leave the Grande Galerie, you are immersed in a comprehensive aesthetic environment. When you walk into the building, you slink into a gloaming. The dusky aquatic lighting of muted neon blue draws you from under a low-ceilinged space into an increasingly open floor plan. You cannot really tell at first where you are or what direction you should move in; if you can give yourself up to the environment, it is quite pleasant and exciting. There are patches of soft blue and green light gliding slowly across the walls and across the people. There is a sound of water quietly rushing all around you, and as you begin to cognize the suspended animal specimens, you come to realize that you are in a virtual aquatic environment. There are sea cows floating above you on some kind of clear glass plates; through the ubiquitous glass walls you see an enormous sperm whale skeleton hovering to the right; there is an oddly placed display of forelimb homologies, but no explanation; there are schools of very realistic fish and dolphins all around you, frozen in mid-dart on thin steel rods or invisible threads; there are display cases, which don't seem at all like cases because they are more like thin, frameless glass walls, that miraculously suspend mollusk shells and various other elements of biotopes within them. These cases have abstract drawings of geological environments etched into them. The etched lines are translucent and meander around the specimens, giving you an intuitive diagrammatic sense of the local habitats of the suspended fauna. There are resting benches everywhere, and just when you're thinking that the French must get winded easily, you realize that all the textual descriptions of the specimens are contained in the benches. The benches contain retractable Plexiglas slates of scientific information, color-coded and numerically linked to the exhibits. Throughout most of the museum, amazingly, there are none of the descriptive placards and paragraphs that usually frame specimens. This curatorial practice, which defrocks the museum of its mediating atmosphere of written text, actually makes the space feel more like a natural environment.

Next you ascend onto a stadium-sized wooden platform that comprises the terrestrial floor (above marine level) of the whole gallery. It feels as though you've been miniaturized and dropped into a huge diorama, because the skylight-filled ceiling now arches three stories above you, and you are surrounded by large mammals. Beautifully mounted elephants, rhinos, zebras, giraffes, and more seem to amble forward, appearing remarkably casual and remarkably free (though the irony of their being taxidermy mounts is not lost).

All around this mammalian migration are transitional creatures that are partly in the marine environment and partly in the terrestrial one. The spaces within the gallery are so open and interconnected that the same animal—say, the sea cow I saw earlier—can be viewed from beneath, then at eye level, then later from above. There are multiple perspectives on every specimen. On one side of the mammal exodus there is a vaulting latticework structure that rises three flights through the central space, containing meticulously prepared taxidermic mounts of birds, lemurs, sloths, and such. To the side of this three-dimensional specimen graph is another of the noncase cases, this one with an etched, and very idealized, drawing of a tropical tree climbing up the giant pane of glass. Inside the glass, positioned in their respective habitat niches like Cartesian coordinates, you find the relevant ground lizards, then, higher up, the various species of butterfly and flying insects, then various bird species. All of these creatures, arranged by ecological parameters instead of morphological similarities, are displayed together. They are integrated rather than being separated into discrete cases for lizards, butterflies, birds, and so on.

The second and third floors are like suspended hallways that frame the expansive nave. The second floor is a thoroughly depressing and heavily moralized tour through the human destruction of nature. Of course, the curators don't explicitly sell the exhibit this way. When you start out, you read this bench slate: "Relations between man and nature: Human activity has transformed the evolution of the natural world, causing rapid, profound, and sometimes irreversible modifications to the environment." This really translates into "Tour the astoundingly many ways that humans rape the planet, and find new and interesting reasons to loathe yourself and your species." Understandably, we all deserve this self-loathing, and it has become the single greatest constant (regardless of nation or state) of the contemporary museum's moral rhetoric. Like Fredrick Ruysch's dioramas, which called upon his eighteenth-century audience to repent their sins because the flesh is ephemeral, contemporary exhibits call their audiences to repent and reverse their industrial ways because the earth is fragile.

The second level details the ways that human wastefulness and exploitation cause far-reaching pollution and the extinction of flora and fauna. Here the gallery, like most other contemporary museums, is not acting as a storehouse for collections, nor as research institution, nor as entertainment. It is trying to make the vitally important, and frequently abstract, idea of ecological connectedness palpable and ethically transformative for the patron. In reality, the term "ethically transformative" is simply too heavy here. No curator believes that patrons will drop to their knees in a moral epiphany. Moral sentiments will not suddenly materialize where there previously were none. But dormant moral sentiments (of responsibility, duty, guilt, compassion) can be rekindled by such museological rhetoric, and exhibit imagery can play a powerful role in this mission.

When I was touring the second level, I sat down in front of some crushing images of factory-farming abuse. I pulled out one of the information slates and read the article titled "*Les rapports homme-animal chez les chasseurs Inuit*" (Relationships between man and animals:

Inuit hunters). This slate was offered as a kind of antidote to First World exploitation. Here I learned that the Inuit of the Arctic regions of Siberia, Alaska, Canada, and Greenland hunt animals with respect. The info slate went on as a paean to the Inuit's symbiotic, "reciprocal," "complementary," "mutual," etc., relationship with animals. The article continued by stating that this healthy relationship was based on important moral ties. Of course, as I read this and looked at the images, every one of my heartstrings was twanging like a penitent banjo.

As you wind your way toward the end of level two there is a gripping little display called, simply, "Pollution." Here stands a chunky block of hard-packed garbage, about the size of a refrigerator. The debris is a combination of commercial plastics, paper products, metals, household detritus, some kind of unidentifiable organic-looking slime, and more. The sign explains (there's actually a sign here) that this evil lump is the genuine rubbish produced by a four-person French family in just ten days. It is a powerful display.

Finally, after a few more exhibits of human idiocy, the entire second level culminates in a little circular room. Inside the little room, you hear droning New Age music and find two monolithic plate-glass structures on either side of a tiny replica of our planet. Etched on the glass are reproductions of primitive cave paintings and various unintelligible but mystically inspired symbols. There is no text anywhere. The little room has a blue-pearl kind of tranquility to it. An intense beam of light strikes the tiny orb, but the whole thing is just too melodramatic. When Heidi and I were there, we spent our time in this histrionic chamber laughing and singing bad versions of Strauss's "Thus Spake Zarathustra."

The top level of the gallery is devoted to the history and theory of evolution. It is psychologically elevating to stroll through this beautiful balcony space, with its precipitous drop-off to the animal exodus below. The ambient quality of the whole museum is darkness, but well-placed lighting punctuates the exhibits effectively and even sculpts open spaces into agreeable quasi-compartments. The first thing that you encounter, on level three, is a weird group of miscellaneous dun-colored creatures hanging from the low balcony ceiling. The creatures are very old-looking mounts and dried preps (desiccated, air-blown, and duly shellacked), and some are misshapen in places and bordering on the fantastical. It is as if the fanciful creatures of Gesner's *Historia animalium* were floating above you. I could find no textual explanation for their existence when I was there, but those in the know could clearly see a wonderful homage to those museum ancestors the curiosity cabinets.

Suddenly eerie music starts to build from nowhere and everywhere. It is not really music per se, but a strange mixture of tempest sounds and guttural harmonies, like a choir of Tibetan monks collapsing into an orchestra pit. The sound becomes increasingly loud and creepy. As you look out over the balcony, you find that the lighting is changing drastically across the mammal pilgrimage. This makes the animals below seem to be moving, and it takes a while to perceive that this is just a trick of the lighting and not robotics. Things are getting darker everywhere, which is alarming, since it is already rather dark. Ominous sonic belches are building up from some state-of-the-art sound system. Kids start to sputter and shake with a tense combination of fear and glee. Suddenly lightning flashes throughout, and rain sounds rise. The museum is simulating a skull-clutching rainstorm, but it's really much weirder. Maybe it's the disturbing Wagner-played-backward music or the intoxicating light show, but it is not like any storm you have ever experienced. It's how a thunderstorm would be if nature were managed by Cirque du Soleil.

After this dies down, you return to the evolution exhibits, where you find small diorama chambers built into the walls. Here are little historical chapters, mostly consisting of books and little decorative fossils, that chronicle the different naturalists. We find window digests of Lamarck, Cuvier, Geoffroy, Darwin, Mendel, and Watson and Crick (but mostly Lamarck, Cuvier, and Geoffroy). Then a corner exhibit seems vaguely familiar. It is the identical display of forelimb homologies that one encounters down near the sea cow, only now there is a context of evolutionary explanation. The image of homologies is repeated as you ascend through the museum, and, as it does so, its significance begins to bubble up to the cognitive level of your brain. When you first encounter this case of skeletal bat's wing, human's arm, mole's claw, sea cow's flipper, and so on, it is just curious and provocative. But now, at the evolution level of the museum, it appears in a meaningful theoretical context. Past this homology display, you find the final exhibits: a series of similar window dioramas on the topics of heredity (titled quaintly "Memories") and genetic mutation (wherein fruit flies abound).

By far, the main rhetorical emphasis of the Grande Galerie de l'évolution is biodiversity. *Biodiversity* has become a fairly common term in the last decade. Once a neologism of technical nomenclature, biodiversity is now an expression that grade-school kids wield with confidence. Simply stated, it means the variety of life on earth, but it effectively suggests much more. Biodiversity is the full range of organic expressions: from blue-green algae to sperm whales, from trilobites to sticklebacks, from humans to slime molds, from deserts to rain forests, from dinosaurs to DNA. Generally speaking, scientists break down this overwhelmingly big idea into three categories that reflect different levels of biological organization. There is biodiversity at the genetic level, the species level, and the ecological level.

Generally speaking, the term *biodiversity* picks out a feature of nature that is at once obvious and overwhelming. Nature's forms radiate out into stupefying variations, veering and swerving in millions of separate trajectories, like myriad sparks from a flame. The Paris museum actually conveys this reality to you. It has no interest in transmitting taxonomic truths to its patrons. Unlike Emerson's experience of Cuvier's tamed nature—a nature made precise, orderly, and lawful—the new Grande Galerie seems to strip away the focusing prism of system and lets nature spill forth with sublime complexity.

The lower and first levels of the Grande Galerie are collectively entitled "Life: An Amazing Spectacle." These are the areas that I described earlier, where one moves through underwater biotopes and then emerges to the mammalian exodus. The museum brochures attempt to give equal weight to the second and third floors ("Man's Primordial Role" and "Evolution," respectively), but anybody who tours the museum knows better. Significantly more space is given over to the biodiversity in "Life: An Amazing Spectacle" than to the other museological themes. And even when you are in the open balconies of the second and third levels, you spend much of your time hanging over the handrails looking at the spectacle down below.

When first entering the oceanic environment of level one, it is difficult to appreciate the cleverness of the exhibit. You are so drawn in by the aqua light and the mounted sharks and fish all around you that you only slowly begin to see that you are actually moving through the biodiversity of different oceanic ecosystems. First you are presented with the pelagic environment (open waters), where all the specimens are aptly suspended without sea-floor support or reef structures or any sustaining geological formations. Plankton, cetaceans, and other organisms seem to float about you. Next you move into a littoral zone; these zones run along coastlines and have various depths. In these zones, where alternating submersion and exposure by tides is normal, you find various sea horses, crustaceans, tidal fish, and seaweeds. The wet-prep specimens are artfully placed in the futuristic glass cases, and there are abstract line drawings and video monitors throughout. After this you move into the coral reefs ecosystem, where one encounters dried coral specimens and wet preps lining the shelves of stylized cabinets. A large walk-around blob of plastic virtual coral sits in the center of the ecosystem. Kids can massage the protuberant bulges and fondle the hairy sprouts that emerge from it. This is followed, at first imperceptibly, by a marine environment of hydrothermal springs. The springs exist because water, seeping into the sea floor, is heated by magma below and gets forced up through pressurized vents. This ecosystem is rich with bacteria and various minerals, and many of the larger organisms are tubular filter feeders. Finally you enter the abyssal zone. This is the ecosystem of deep ocean trenches. Icy cold waters, very high pressures, and impenetrable darkness make this an environment for only a few tiny spiderlike invertebrates with filamentous bodies, and gigantic nightmarish squid.

The specimen placement in the biodiversity sections carefully employs subtle visual symbolism as well. For example, in the display windows of the abyssal zone, the specimens are placed very far apart from each other. These relatively sparse cabinets (almost empty) are visual representations of the rarity of life in these regions. In contrast to these cases, one finds the specimens of the hydrothermal vents section clustered in tight, busy groups. Using spatial proximity and distance, the curators can impart truths about biodiversity, and in this nondiscursive format you may not be entirely cognizant of your newly acquired knowledge.

As you move upstairs and out of the marine diversity exhibits, you encounter a similar combination of dizzying colors, shapes, and sizes. The cool blue and green lighting gives way to warmer earth tones of brown, yellow-orange, taupe, and so on. The warm-blooded mammals of level two are accentuated, and contrasted with the briny deep, by sunny gel-filtered spotlights. This time you are on dry land, but again the exhibits merge almost imperceptibly into diverse species and diverse ecosystems. First you move through the African savanna, where a multitude of autochthonous plants lines the march of herbivores and predators. Many cabinets along the way give you the impression of tremendous insect diversity and density. Next there is an equally elaborate stroll through the tropical forests of the Americas; then the utterly extreme flora and fauna of the Sahara; after that the animals and plants of the Arctic and Antarctic are explored, and finally the local fauna and flora of France (a temperate region that houses some five hundred species of vertebrates and eight thousand species of flora). And if you are too overwhelmed to pick up on the subtle museological message, there is also a fifteen-minute film about biodiversity.

* * *

The French see nature as a sprawling, uncontainable spectacle, and one can truly comprehend this fact when the Grande Galerie is compared with New York's American Museum of Natural History. The AMNH continues to stress taxonomy; it wants its patrons to appreciate the *order* of nature. Of course, taxonomy has had some interesting developments since we left off our story in the nineteenth century. A few of them need to be noted briefly if we are to understand the way that the AMNH constructs nature.

Given that Darwin's mechanism of natural selection describes an engine room of adaptation, it would seem that Darwin himself would have followed the functionalist approach to taxonomy, grouping animals with a Cuvierian focus on adaptation. Interestingly enough, while Darwin stressed the role of natural selection in shaping the specific adaptations of a species, he argued that the classification of organisms (the relationships they have to each other) should not be based on adaptations. Grouping animals that have adapted to life in the air, for example, would be artificial; it would not reveal the natural system of classification. Darwin resuscitates talk of a "natural system" of taxonomy, but the term has been redefined. In the seventeenth and eighteenth centuries the term had theological implications, but for Darwin (and after) "natural system" became synonymous with "genealogical system." In the *Origin of Species* he referred to genealogy as the "hidden bond which naturalists have been unconsciously seeking" in their classification practices.

The natural or the correct way of establishing taxa was by discovering the historical relationships between organisms: Who is related to whom, and by what common ancestor? Taxonomy based on adaptations alone, such as similar traits for living underwater, will not reveal lines of descent. Natural selection can give cetaceans and fish some very similar adaptations for aquatic life, but these do not reflect close ancestry. Unrelated animals can converge, over long periods of time, on similar adaptations (similar solutions to common environments), and the taxonomist must be wary of confusing convergent likeness with family resemblance. The way to avoid this possible confusion, Darwin argued, was to be pluralistic in choosing characters for comparison. Unlike the Linnean method of choosing one or two characters for comparison, such as feeding organs or sexual structures, Darwin thought that comparing many characters would be more likely to spot true lines of descent. In the *Descent of Man* he claims that "numerous points of resemblance are of much more importance than the amount of similarity or dissimilarity in a few points." Moreover, Darwin completely redirects taxonomy away from the functionalism of Hunter, Cuvier, and their colleagues when he argues that "resemblances in several unimportant structures, in useless and rudimentary organs, or not now functionally active, or in an embryological condition, are by far the most serviceable for classification; for they can hardly be due to adaptations within a late period; and thus they reveal the old lines of descent or of true affinity." This historical interpretation of taxa has been struggling for popular comprehension since Darwin's day. Contemporary taxonomy (or systematics) is still frequently misunderstood by the general public, and no museum is trying harder to clarify this than the AMNH.

In June 1996 the American Museum of Natural History opened its Hall of Vertebrate Origins. This hall, which is touted as "the world's most comprehensive exhibition of vertebrate fossils," is part of a larger exhibit circuit that includes a hall of dinosaurs and a hall of mammals. The exhibition space and overall aesthetic are very different from the Grande Galerie. Instead of the enveloping and mysterious darkness of the Paris gallery, New York's space is flooded with light. The luminous gallery space is like a visual representation of enlightenment itself, with knowledge and rationality seeming to blast out from windows high above, spotlights on the ceiling, and Mission-style hanging lamps. The neoclassical walls, arches, and moldings have been beautifully restored and form an elegant wedding-cake white backdrop for the richly dark fossils (fig. 5.1). But apart from these aesthetic differences with Paris, some major conceptual differences can be detected as well. The AMNH space has been organized by curators into a neat geometry of branching pathways. The spatial organization and specimen placement embody a specific theory of taxonomy (systematics) called cladistics. Very few individuals from the general public have ever heard of cladistics, yet it is the intellectual foundation for the entire AMNH exhibit. The New York curators are pioneering a new rhetoric of museum display.

In the 1950s two new schools of systematics began to emerge and battle for dominance: cladistics (the name is from the Greek term for "twig," *klados*) and phenetics (the term stems from the Greek *phainein*, "to show or appear," sharing the Greek root with *phenomenalism*). These two different schools disagreed strongly about the method for organizing taxa.

Following Darwin, most taxonomists of the first half of the twentieth century were isolating homologous structures between animals and grouping these into historical lineages (phylogenies). But around the middle of this century some taxonomists were getting nervous about this methodology and general objective. To some taxonomists the method seemed circular, because in order to recognize

Figure 5.1. A ray-finned fish (actinopterygian) hangs from the ceiling of the Hall of Vertebrate Origins, American Museum of Natural History, New York.

a homology (as opposed to a convergent similarity) you had to infer a phylogenetic (historical) relation; but in order to establish a phylogenetic relation, you had to recognize homologies. Some taxonomists were nervous that a theory (evolutionary taxonomy) was being confirmed by observations (morphological similarities) whose significance depended upon the assumption of that very theory. How does the taxonomist know, for example, that similarities between a handful of characters (say, bushy tail, dentition, and flight-enabling skin flap) will produce a genealogical taxon (such as flying squirrels) rather than an artificial grouping of convergent traits (such as placental flying squirrels and marsupial phalangers)?

In response to such worries the pheneticists decided to give up the claim that classification revealed evolutionary (historical) lineages. They decided to drop out any theory whatsoever, claiming that evolution should not be assumed before the observations. The classification of organisms, they argued, should be pure and theory-free. This was not because of some hostility toward evolution; the pheneticists believed, like other biologists, in evolution. They simply did not believe that one should *begin* to classify organisms by assuming evolutionary connections.

The pheneticists were linked to phenomenalist ways of thinking by more than their name. Remember that the phenomenalist naturalists of the Enlightenment period (people such as Ray, Buffon, and even Hunter and Cuvier to some extent) held that knowledge comes through the senses. This being so, they argued that our knowledge of reality is, to be painfully precise, only a knowledge of the appearances of reality. For the phenomenalist, our reality is always virtual, because our senses mediate the world for us. Some naturalists and philosophers believed that there was a reality (God, atomic agitations, etc.) beyond the phenomena, a reality that caused the appearances. And other phenomenalists quickly lost interest altogether in a realm that was, in principle, incapable of being accessed or talked about. In the early part of the twentieth century a new group of these phenomenalists, called logical positivists, reiterated this philosophy.

The positivism of the twentieth century grew out of the intellectual environment of Vienna in the 1920s and '30s and included such influential characters as Karl Popper, Ludwig Wittgenstein, and Rudolf Carnap. Positivism held that scientific propositions were meaningful because they referred to verifiable experiences. Following the empiricism of David Hume's much earlier work, the positivists claimed that meaningful claims and theories were shorthand summaries of sensory experiences. Gravitation theory, for example, is a linguistic-mathematical representation of what an observer sees when experimenting on moving bodies. Theories (and sentences, for that matter) are meaningful to the extent that they refer to possible or actual sensory experiences.

The pheneticists adopted a positivist attitude toward their work. They argued that the individual organism and the taxon were both rational constructions of sensory data. As scientists, we take different sense data, such as colors and shapes, and clip them together with a classificatory concept or designation such as "monkey" or "rodent" or "*Sciurus carolinensis*." The objective foundation of the taxonomist's science is nothing but those sense data; the classificatory character is itself a particular construction of sense data. Scientists, much as they might like to, cannot access some more ultimate reality beyond these sense experiences. And like John Ray's much earlier skepticism about finding the one or two characters that would create a "correspondence" between human taxonomy and God's taxonomy, pheneticists became skeptical about using only a few characters.

Pheneticists saw their theory-neutral approach as starting from scratch, without assumed theories about which characters to emphasize and which to ignore. The most objective way to build a taxonomy, they argued, was to measure overall similarities rather than privilege a few characters. This whole approach to taxonomy, charting every part of every organism and calculating overall matches, would have seemed ridiculous from the practical viewpoint, except for the new tools of computer technology. Computers seemed, to the pheneticists, to be the perfect way to calculate huge amounts of data and arrive at unbiased matchings of characters. Without weighting some characters as more important than others, the computers might be able to arrive at a numerically based taxonomy. Each of the many characters of one animal contain different states (e.g., present or absent), which are given numerical values (and which get crunched into a measure of overall similarity), and by using mathematical algorithms the species and other taxa are clustered according to shared overall resemblances. These clusters may or may not reflect phylogenetic relationships between taxa, but the pheneticists don't know and don't care. The phenetic approach refuses to infer unobservable truths. This kind of taxonomy, staying true to its phenomenalism, does not give us any information beyond the interplay of appearances; all it provides is a map, called a phenogram, of sense-data clusters.

Phenetics seems, at first, like a really good idea. Who doesn't want to expunge theoretical biases from science? But by the mid-1970s phenetics was moribund, and in the '80s it was well and truly dead. Its competitor, cladistics, has become *the* way to do taxonomy.

Phenetics falls apart for a number of reasons, but perhaps the most striking is that no one knows what "overall similarity" is. Also, it is not really possible to pretend that one can be entirely theory-neutral. In addition, phenetics uses negative characters (e.g., an extreme example would be "lack of vertebrae" for Invertebrata). Cladists, by contrast, would recognize "presence of vertebrae" as a character of Vertebrata only when it is congruent with other empirical data. The problem with negative data for an organism is that it is effectively infinite, and this is a major factor in why one cannot determine overall similarity.

The idea of overall similarity seems intuitively clear at first, but on closer examination it breaks down. Without first weighting some characters over others, computers will evaluate all traits equally. The size of the animal will be weighed equally with the heart structure, the color of the nose equally with the respiration equipment, the shape of the tail equally with the shape of the colon, and so on. But it seems impossible to itemize an animal in a completely theory-free way. Without some theoretical guidance, we would not even be able to determine at which phases of life we should collect our sense-data characters. In addition to the problem of how one unarbitrarily fastens a numerical property to qualitative observations, one is left unclear on whether to gather the data from the creature's embryonic phase, larval stage, pupal stage, or adult stage. Or, if all of these, how does the numerical integration occur for one organism, let alone for all organisms?

Systematist and philosopher Michael Ghiselin uses an interesting analogy to isolate the ambiguity of overall similarity. He points out that the idea of overall similarity makes sense to our rough-and-ready practical minds when we reflect on the fact that we can easily place together apples, oranges, bananas, and such in a supermarket. There seem to be obvious properties that each of these things possesses, but it all gets quite complicated very quickly when we start factoring in many traits (and when we leave behind the conventions to which we've grown accustomed). For example, supermarkets place canned fruits and vegetables in a different group from fresh fruits and vegetables. Ghiselin points out that "once one tries to decide whether a canned peach is more similar to a fresh mango or a canned tomato, one begins to have difficulties, and not just because the tomato is botanically a fruit and gastronomically a vegetable." For taxonomy of any kind to work, one must stipulate before the analysis (at least provisionally) that some limited set of properties will be compared. Similarity is always relative to those stipulated properties.

The cladistic school of taxonomy, which outlived phenetics and became the basis for the floor plan of the AMNH, takes a very different approach. Cladistics attempts to group taxa hierarchically, based on shared derived characters (hypotheses of homology). Unlike the pheneticists, many cladists interpret their classification patterns to represent genealogical relationships. All the members of a taxon must have a common ancestor. The clustering of organisms into subordinate taxa such as classes or genera reflect the evolutionary relationships of those taxa. In order to organize biodiversity data, the cladists weight characters according to the idea of primitive and derived structures. But, again, the terms *primitive* and *derived* are employed only as relative designations. Cladists try to find the most general taxonomic level for each unique character. The term *primitive* is not used by cladists to refer to a particular locus in the phylogeny of all life.

When you enter the Hall of Vertebrates, you first encounter a tall display column that reads "Jaws." Vertebrates with jaws are called gnathostomes. Looming over this column is a huge set of toothy shark jaws, strangely abstracted from their fifty-foot *Megalodon* owner. On the column is a series of stipple-inked line drawings of different animal skull profiles, a clinical fashion show of different jaw structures. Lower on the column you are informed about the material that makes up the jaw ("special cartilaginous structures, usually encased in bone"); then the anatomical relation between the jaw character and its surrounding structures ("in primitive gnathostomes, the upper jaw is hinged and movable, with joints or ligaments connecting it to the skull"—think of sharks here—whereas in more advanced forms "the upper jaw may be fused to the braincase"—think of yourself here); finally, this information is followed by a description of some of the functions of the jaw ("ability to bite," "assist in food capture," "crushing ability," "food filtering," "venom delivering," "gill respiring," etc.). You are standing, here at the "Jaws" column, on an early evolutionary branching point. You are at the base of a vertebrate cladogram, and each successive character column that lines the central axis of the gallery is another branching point.

The central axis of the gallery represents the trunk of the evolutionary tree, and it is punctuated by those successive character changes that mark the origins of major taxonomic groups. Along this central axis are little branches that diverge from either side of the trunk, and these represent the different taxa that still share the major character. For example, the major character "jaw" drops a branching point that creates two very large categories of vertebrates: those with a jaw and those without. But many diverse taxa are nested within these major divisions. Living lampreys and slime hags, for example, branch within the jawless category, and extinct placoderms, living sharks, and many others branch within the jawed category. So as you move to one side of the central corridor of the gallery, you encounter giant armored fish suspended above (with corresponding displays below); moving to the other side of this vertebrate gallery path, you encounter sharks, rays, and their relatives the chondrichthians.

The next branching point in the gallery represents the evolution of tetrapod (four-limbed) forms, where you encounter frogs, salamanders, lizards, and so on. Then further up (and remember that you are already in the clade of tetrapods now) you come to the branch of amniotes (animals that reproduce using watertight eggs, and these can be either oviparous or viviparous). Then, as you move along inside this category of amniotes, you will eventually find more branches within branches. And so you continue to wend your way through this new taxonomic logic.

Each nested clade (group) includes an ancestor, all of its descendants, and only its descendants. So the giant clade of vertebrates contains the original vertebrate ancestor (something like a slime hag) and all living and extinct vertebrate forms, but no invertebrate forms. The clade inside the vertebrate group known as tetrapods contains the original four-limbed ancestor (something like the aquatic *Acanthostega*) and all the living and extinct four-limbed descendants.

As you move through this museum space, you slowly begin to appreciate that it is not quite like the usual evolution exhibit. In some sense you are walking through time as you progress through the gallery, but in another sense you are just moving through logical space. Most evolutionary exhibits, such as the Field Museum's "Life over Time," are set up as linear promenades through time, wherein the first displays (the beginning of the walking path) are extinct forms and the later displays (further along on the path) are living forms. But here at the AMNH, the first step on the promenade, long before you even reach the extinct pterodactyls, contains contemporary sharks. Why is there a living animal placed before an extinct one?

Moving through the cladistic gallery is moving from primitive to advanced, but this does not always mean that you are moving from extinct ancient forms to contemporary forms. Moving forward along the axis is actually moving inward into increasingly less populated clades—if you will, into increasingly smaller nested Russian boxes of classification. Inside the big box called Vertebrata is a smaller box called Tetrapoda (four-limbed creatures), and inside that one are two smaller boxes, one called Synapsida (which contains smaller boxes, such as Mammalia) and one called Reptilia (which contains smaller boxes, such as Aves). In this case, we can see the relativity of the primitive/derived distinction. "Presence of vertebrae," for example, is a primitive character within Mammalia, but a derived character for Vertebrata. Moving from the beginning of the Hall of Vertebrate Origins to the end is like moving from the bottom of a cladogram to the top. The branching points near the top represent more recently evolved characters, but this does not mean that the same applies to the organisms themselves. Organisms are complicated mosaics of both primitive and derived characters. Obviously, the sharks or the coelacanths, for example, are currently living forms, but their primitive characters place them low on the cladogram. Cladograms show us a *logic* of life that actually emerges out of the *history* of life. For example, given what we know about phylogeny, organisms need to be vertebrates in order to be milk-producing mammals, but this necessity does not run in reverse (you do not have to be a milk producer to be a vertebrate). People who are accustomed to the exceptionless laws of physics won't be very impressed by this sort of biological generalization. But the subject matter of biology is historical, which means that it matters to the understanding of such entities *where* they are located in space and *when* they are located in time.

Some biologists, most notably Ernst Mayr, have raised objections to the cladistic approach by pointing out its overly syncretic vision.

Cladistics seeks to link together taxa by focusing on their shared characters (the markers of genealogical connection). Stressing only common ancestor relations is not very helpful, however, to the biologist who wants to understand how and why animals differ. Once cladists have established a branching point (say, the common character "openings in the roof of the mouth," which creates a clade of birds and reptiles known as sauropsids) they see their job as pretty much finished. But Mayr points out that many biologists are interested in the subsequent fate of each branch. Once a sister group splits off and occupies a radically new niche (as protobirds did by entering the skies), then specific self-contained traits can create a population very, very different from its nearest clade companion (in this case, the crocodile). Following out an animal's specialized adaptations (rather than their common characters) can lead to important information about a particular origin scenario, transformation, or functional implication, and taxonomy should help generate these diversity-based insights. Some of these criticisms of Mayr's recently have been incorporated into an improved theory and practice of cladistics. But whatever small-scale criticisms and anomalies may continue to crop up for cladistics, it is very securely installed in the current seat of theoretical privilege.

This is a good point at which to reopen briefly the perennial debate between taxonomic skeptics and taxonomic realists, a debate that wanders like a fault line through our entire story of the taxonomic intoxication. Museums, and collections generally, are ordering systems. And while no one believes that museums are perfectly objective mirrors of nature, there is still a question as to whether some taxonomies are better than others. A reasonable skepticism (some might say relativism) seems to follow from the repeated episodes in the history of science in which our brazen confidence about the true structure of nature is deflated. Some intellectuals have looked at the anomalies in taxonomy, the subjectivity inherent in any ordering practice, the comparative differences of other cultures' classification systems, and the logical possibility of ordering things differently; and have marshaled all this as evidence that there is no such thing as objective classification.

Contemporary biologists, not too pleased to hear that they've been spinning their wheels, have a very thoughtful response to this relativist critique. The reason why many anthropologists, poststructuralists, and even playful literati (remember Jorge Luis Borges) have suggested that *all* ways of classifying are thoroughly relative is because they have mistakenly assumed that *logical* classifying is the only kind of classifying. Contemporary systems of classification, based on evolution, are very different from their Linnean ancestors, because evolutionary classification is *not* another logical tree of relations. It is a historical tree of relations, and this makes all the difference. Evolutionary classification does not designate similar objects together into categories; it actually picks out temporally and spatially extended individuals. It reconstructs the history of a changing individual, only these are not the individuals that we're accustomed to. These individuals are huge populations that spread over large expanses of space and time. Such family trees are historical entities, yet they are still more like individuals than classes. Reconstructing a historical narrative is not entirely objective, but such a narrative does not try to group events together because they have outwardly similar features. It groups them together because earlier ones gave rise to specific later ones—the affinities are only by-products of that true "hidden bond." The affinities themselves, while they help scientists to recognize phylogenies, do not create or define the phylogenies. Compare this with a nonhistorical form of classification that actually creates classes by stipulating pass-fail defining criteria. Each member of such a class will have some formal relation with the defining criteria (i.e., it instantiates the concept), but those members will have no significant relation to each other. Think, for example, about the category "paintings." We have some defining criteria (one of which might be something like "coated surfaces carrying designs in pigment"), and this creates a class whose members (of which there is a staggeringly large number) have nothing but the criteria in common. You can see how things really are quite relative in this kind of classification. But it is totally different in contemporary biological classification, where the members are members by dint of their being related to each other.

I can create a logical tree of musical instruments that divides into wind instruments, string instruments, percussion instruments, and so on, and then further divide the wind instruments into reeds and nonreeds. Then reeds can be divided into oboes, saxophones, and the like; then saxes can be separated into tenor, alto, baritone, and so on. All of this seems quite logical, but it is not genealogical. Perhaps an equally possible classing could be to begin with the class of musical instruments and then divide them into metal instruments and wooden instruments; then shiny and nonshiny instruments; then golden-colored, silver-colored, and brown-colored; and so on. Here poststructuralists and other relativists get very excited and claim that evolutionary taxonomy is ultimately no different from these internally consistent but ultimately arbitrary systems. Evolutionary taxonomy *is* different, but it's hard to see why, because it, too, iterates divisions based on similarities and differences. But a historical tree of descent is unlike any other classification—in fact, it's not properly speaking a classification but rather the cause and explanation of a classification. Think about this analogy. If a forest preserve park is host to three family reunions one Sunday afternoon, then the affinity of the Smiths' chins and the affinity of the Johnsons' eyebrows and the affinity of the Coopers' noses may allow us to discriminate between and demarcate the groups. But those affinities are completely unlike the silver and gold colors of the musical instruments. It is the real historical descent of Grandma Smith and Baby Smith that makes the chin-chin affinity. The affinity itself does not bond them together in the way that the gold color supposedly bonds certain instruments into an artificial group.

Granted that there is this all-important difference between phylogenetic classification and other forms of classification, the cladists must still respond to some needling challenges. Phenetics may be dead, but its brief life span did help to raise some important and lasting questions for all taxonomists. How do we determine whether a specific character is relatively primitive or derived? How can we distinguish between homologous and analogous structures? And how can we avoid a tautologous circularity of theory and observation?

Some systematists, including both cladists and pheneticists, remain so resolutely phenomenalist in their approach that they are uninterested in any evolutionary inferences. Cladists and pheneticists are really just trying to construct the most parsimonious arrangement of empirical data. But many evolution-minded biologists seek to establish the genealogical underpinnings for these parsimonious arrangements. If we take a pluralistic approach, incorporating aspects of the different systematics schools and evolutionary biology, we can try to reconstruct the evolutionary trajectory of certain organic structures.

If we are examining the hearts of certain animals, for example, we encounter different heart structures. Some animals have two-chambered hearts, while others have four-chambered ones. Which of these structures, the two- or four-chambered heart, evolved first? The fact that one is more complex than the other is not a good rationale for believing that it is younger, since there is abundant evidence of simpler structures arising after complex structures. Some romantic relativists get excited about this sort of epistemic challenge. But biology rises to the challenge by offering up three strategies for resolving such puzzles. First, one can examine the fossil record, where clear evidence might be found for comparative dating of certain characters (this works better with hard parts instead of soft tissue such as a heart). Next, one can turn to outgroup comparisons. This is when one tries to establish which character is most widely distributed among groups of animals other than the ones currently under study. If many more outgroup taxa have the two-chambered heart than the four-chambered one, then it suggests that the four-chambered heart is probably a later specialization. Last, one can study the embryological development of the organisms to see if the heart starts out as a two-chambered structure and then develops into a four-chambered structure, or vice versa. There is reason to believe that such embryological developments may be repeating phylogenetic developments—that is, the developmental stages of ancestral forms can be seen in the embryogenesis of modern forms. Evolutionary biologist Leonard Radinsky gives an example of this embryological method:

> A classic example of the value of this approach is the demonstration from developmental series that two of the tiny bones in the middle ear of mammals (the malleus and the incus) were derived from the jaw bones (articular and quadrate) of our reptilian ancestors. That is, the early developmental stages of the ear bones in mammals resemble early stages of jaw development in reptiles. This inferred evolutionary transformation of reptile jaw bones to mammal ear bones was later confirmed with the discovery of "missing link" fossils that display the intermediate condition in adult stages.

Next is the challenge of homology versus convergence. How do we determine whether the affinities we observe are signs of common ancestry, and not merely products of convergent evolution? This was a paralyzing difficulty until the very recent development of biochemical comparative technology. Many contemporary systematists, including Dr. John Bates from the Field Museum (chapter 1), use a comparative technique called DNA hybridization. Biologists can compare the DNA of two species by separating and then fusing DNA strands. If you wanted to compare starlings and mockingbirds, for example, you would heat starling DNA until it separates out into single strands, and you would do the same with mockingbird DNA. When you combine single strands from each bird's DNA, the two strands bond together, but the bond is not as tight as the double helix of a single species, because the two strands are not a perfect match. Next, you heat the DNA hybrid and measure the precise temperature at which the bond breaks and the strands separate again. When two species are genetically very close they share many DNA bonds, so it requires higher temperatures to separate the strands; lower temperatures indicate greater genealogical distance. All of this is painstakingly calibrated with molecular-clock rates of change, and the result is an independent checking system for corroborating claims of genealogical connection. I suppose a relativist, truly bent on radical skepticism, could counter this solution by claiming that we have the homology-versus-convergence problem all over again at the molecular level. In other words, if convergence can deceive us by creating similar bodies for unrelated organisms, why not similar DNA structures for unrelated organisms? This is maddening, but it has to be acknowledged. It is unclear at this time whether such a concern will become an annoying anomaly, will go away quietly, or can be theoretically countered in due time.

Finally, we have the concern that the pheneticists raised about keeping our taxonomy theory-neutral. What can systematists say in response to the charge that they might be letting their theoretical commitments to evolution color or guide their observations? While this seems at first blush like a completely legitimate charge, it's actually a product of an outmoded model of science. We previously believed that scientists just looked really hard at the phenomenon they wanted to understand, and by pure observation alone objective truths would emerge. In some sense this is accurate, but scientists are not, and never have been, theory-free spectators; they can't be, because without expectations or background beliefs or assumptions, every one of their sensory experiences would be equally significant—which means that the experiences would be equally insignificant. The act of observation itself is the picking out of an isolated event or thing from the blooming, buzzing continuum of sense experience. The picking-out process—say, of a species or a clade—is a product of our careful attention to nature's structure (given in sense experience) *and* the specific hypotheses that are always directing our gaze. The objective for most scientists is not to be theory-free, but to be theory-minimal in the empirical phase of their research.

Doing science, according to my old teacher Harold I. Brown, is more like doing cryptography. A cryptographer attempts to decode cipher texts. "She cannot begin until she adopts some hypotheses about the significance of the marks on the paper before her, and her attempts to interpret the text are guided by these hypotheses. But she cannot accept any hypothesis whatsoever [as the relativist wants to claim], because there is still a text before her, and the task is to decode that text." This is not a crippling circularity; it is science as usual.

As you move about in the cladistic galleries of the AMNH, you repeatedly encounter plaques that acknowledge the fallibility of the ordering system before you. You are reminded, with refreshing honesty, of the fact that these specimen organizations are not perfect, nor are they certain. Time and again the AMNH admits that the ordering of organisms is highly interpretive work and that there are inevitable differences between scholars. One cladistic hierarchy is suggested by focusing on one set of characters, and another hierarchy is suggested by other characters. The exhibit catalog defends the displays unapologetically by pointing out that in cases of taxonomic dispute, "the hierarchy consistent with the most features [characters], and tested by many different scientists, has been presented as the most likely." The idea that theory and observation are always influencing each other and that different methodological starting points (character choices) lead to somewhat different groupings is not an indictment of taxonomic science. The catalog's explanation of the AMNH's organizational plan concludes with an open admission of the provisional status of the museum's displays: "The view of evolutionary history seen in these halls represents the best interpretation of the available evidence according to researchers at the American Museum of Natural History. These views, like all scientific ideas, are subject to change and refinement. Further research and the discovery of new fossils may well modify our present understanding."

While cladists generally admit to the fallibility of their classifications, they are quick to close ranks (and are partly correct in doing so) when relativists start poking about. When some deconstructionist wag comes along, claiming that contemporary ordering systems are no better or worse than the medieval spiritualized taxonomies, contemporary scientists are thrown into convulsive fits about the depravity and injurious state of today's humanities. *Relativism* is an inflammatory word, and scientists get nervous around it because it is often used as a synonym for "fabricated truths" or "subjective ideology." Some humanities scholars have chosen to reconceptualize the sciences (and their findings) from the unconstrained perspective of literary criticism. Everything is a text, and everything is political.

If we can move beyond this draconian version of relativism (and for the still-bristling scientific community, this will probably take a while), we can actually see a deeper strata of healthy relativism. Some practicing taxonomists, both today and in Linnaeus's day, are realists in the sense that they see their classification systems as maps of reality. The maps correspond to the real. But there is a distinct difference between the way that contemporary and traditional taxonomists think about the real. The old school, recall, thought that taxonomic realism meant that ultimately there was (or will be) one unambiguously correct taxonomic map of reality. We as investigators may be currently confused, but nature (being a production of the Deity's unambiguous mind) is not. All taxonomic problems, even the mundane ones such as "Where the hell do we place the platypus?" will eventually resolve into a coherent, beatific system. But none of today's working taxonomists, who are still realists in the sense that their maps correspond to chunks of reality, is the least bit interested in this ideal unified map. For them, there is no taxonomy "in itself."

In the current scientific atmosphere, no one is very bothered by the intermingling of natural and artificial classifications. "Natural system," for many taxonomists, means "evolutionary," but very good biology can be done by grouping organisms in a nonevolutionary way. Kingdom-level classifications according to ecological functions, such as producers (plants), consumers (animals), and decomposers (fungi), are acceptable categories for certain kinds of biological research. And while a genealogical classification would never group bats and birds together, a researcher might make good use of such an artificial category when trying to analyze the biomechanics of flight. There are many legitimate ways of dividing up the world into categories. This is *not* relativism, in one sense, because there are objective constraints on the process. But it *is* relativism in the sense that the correct way of grouping things is relative to the kinds of questions one is asking of nature. It is strange to assume that the map of groupings that tells us how organisms are genealogically related to each other will be consistent somehow with the many other kinds of groupings that we find useful. Perhaps there can be no objective univocal taxonomy—no final theoretical map of the world—because the maps will always be relative to our different scientific motives. In fact, the idea that an ordering system could be true independent of anyone's use for or research interest in it is quite possibly just a vestigial leftover of deist thinking. The validity of an ordering system cannot be evaluated in any absolute terms (thinking it can is just pining for a God's-eye view of nature); the system must be assessed relative to the goals and motivations of the researchers. Here is where the relativist view of science history seems rather convincing. It is somewhat muddle-headed to look back, from the perspective of our genealogy-based museums, at Hunter's and Cuvier's museums and consider them wrong. If Hunter was trying to show us the *historical* connections between organisms, then yes, his groupings would be erroneous. But he wasn't trying to do this. We now know more than ever about the historical blood ties between organisms (the goal of evolutionary taxonomy), but we would be obtuse to think that all previous organizational systems were just bad tries (or near misses) at this same taxonomic goal.

Museum researcher Dr. John Wagner explained to me the way all this theory gets expressed in practice. As a practicing entomologist who works with the collections at the Field Museum, he understands the theory/praxis dynamic of a research program.

"In the back of your mind," John said, "you know that the ultimate goal of your work is to figure out evolutionary relationships between your specimens. But the day-to-day work is dominated by much smaller puzzles, and you're gathering comparative information from a wide variety of research tools, like morphology, ecology, and biochemistry. There is a pluralistic use of methodology in an attempt to construct a fuller picture of your organism.

"The museum is a repository for donated collections that people have spent their lives amassing. We just had a donation from Colorado, for example, where a man spent his whole life collecting ants, and before he died he wanted to make sure that his work would find a good home. Universities that he looked into were not committed to taking care of the collection, so he settled on the Field, where we are currently cataloging the specimens. Now, if you're a curator who works on ants and you think you may have come across a new species, you check the literature to see if it has already been described and then you plow through the collections to examine type specimens. You have to work with the collections directly to find the original series or individual ant, because the inherently inadequate text descriptions won't allow you to unravel the puzzle. It's a visual thing, in large part. But more recently the method of DNA sequencing has risen to the top of the methodological pile.

"When I was in graduate school," John explained, "my friend would tease me about my interest in insect anatomy. He'd say, 'Why are you wasting your time with this anatomy stuff? Just get a bunch of bugs, put them between two pieces of filter paper, and drop a heavy book on them. Then you'll have a blot and you can analyze the proteins, and that's all you need.' To some extent his joke has actually come to pass. But science is no different than other walks of life in terms of faddism. When we get new toys to play with, everything moves in that direction. With regard to a specific insect, I'll start out with information from biogeography and morphology and get a certain picture that way. Then I'll drop in the genetics and biochemistry and hope that information confirms the picture that morphology and biogeography provided. In some cases it does. In other cases, it just blows you out of the water and you have to go back to square one and rework it. That's what makes it so much fun. That's why people do it."

With this backstage update of the current skepticism-relativism-realism debates in mind, we can return to the variations of the museums' front-stage displays. We have seen that the Grande Galerie, in Paris, is primarily concerned with a portrayal of nature's diversity—almost a return to the overwhelming aesthetic of the curiosity cabinets. Very recently the AMNH has opened its own version of the biodiversity cabinet in its lower-level space, an impressive explosion of forms called the Hall of Biodiversity (figs. 5.2 and 5.3). It is rich,

Figure 5.2. The American Museum of Natural History's Hall of Biodiversity. The crowded aesthetic emphasis is very reminiscent of the Grande Galerie de l'évolution in Paris.

Figure 5.3. A striking backlit display from the American Museum of Natural History's Hall of Biodiversity.

luxuriant, and very derivative of the Paris museum's aesthetic (and the heavy moral tone of ecofragility is here as well). But the AMNH evolution galleries have just the opposite tendency. Variation, divergence, and the kaleidoscopic profusion of nature are all contracted inside clear taxonomic parameters. The patron of the AMNH primarily sees those things that organisms have in common. The AMNH brochure is also quite clear on this, pointing out that "at first glance, the diversity of life on Earth may seem overwhelming. But a pattern emerges if we look for features that different animals have in common." Nature is girded by common structures. And aesthetically speaking, even the natural differentiations of coloration (fur, feathers, skin, etc.) are stripped away, since all the animals in these galleries are presented as achromatic skeletons (there is no taxidermy). All this clarity and orderliness might seem, in my description, less beautiful than the elaborate biotopes of Paris. But the AMNH's gallery, with its crisp symmetry, is remarkably elegant. Aesthetically speaking, the AMNH is Apollonian and the Grande Galerie is Dionysian.

In addition to these constructions and representations of nature, the museums reveal some interesting national tendencies as well. There is something sort of unsurprising about the fact that the French museum is all about beauty and the American one is all about order. Some stereotypes are in full swing here, I guess. But rather less obvious and more complex interminglings of national identity, scientific theory, and museological practices can be excavated if you compare the Paris museum and the Natural History Museum, in London.

The Natural History Museum, in the South Kensington district, started out as a department of the British Museum in Bloomsbury. In 1753 the wealthy physician Sir Hans Sloane died and left his collection to the British people. Sloane's extensive private collection became the basis for the quickly growing British Museum. In the first half century of its history the museum's collection grew by leaps and bounds, fueled generally by the collecting mania set off by European encounters with new worlds and specifically by Joseph Banks and Captain Cook. Banks, remember, was a good friend of John Hunter and eventually a president of the Royal Society. He always brought skillful illustrators and collectors along on his expeditions in order to record the natural history of the explored regions. These specimens and drawings helped build up the British Museum's holdings. By the 1860s, however, the museum could no longer contain the overflowing specimens. Richard Owen left the curator position at the Hunterian Museum and guided the move of the British

Museum to the new South Kensington institution, which eventually opened in 1881 (Fig. 5.4). In 1963 the Natural History Museum broke its last administrative ties to the British Museum, and it stands today as an independent institution.

In the museum's exhibit entitled "Origin of Species," there is a very different approach to the presentation of evolution. It is not surprising that Darwin should be the sunlike center of the exhibit, around which every other fact and display revolves. He is, after all, a deserving countryman. An almost juvenile national pride can be seen in London's celebration of Darwin—and Paris's attempts to conceal him. But it is not enough, however, to see this as just a story of allegiance to countrymen. The deeper rhetoric of museology reflects longstanding battles over what constitutes good science and which mechanism of transformation, natural selection or mutation, is more important.

The London exhibit begins with a life-sized diorama of Charles Darwin reading in his study. Here you are introduced to the man and his legendary book. The Darwin manikin is supposed to look like a marble sculpture, I think, because he is completely chalk white. He sits, vaguely regal, in an overstuffed reading chair, like a papier-mâché effigy. Past this introductory display, the rest of the entire exhibit is laid out as an embodiment of Darwin's ideas. The many evolutionary thinkers who preceded Darwin are not acknowledged. The story is told in ten sections: (1) the puzzle that Charles faced, and the clue of domestic selection; (2, 3) what is a species; (4–8) the steps in Charles's argument; (9, 10) how natural selection forms new species. The unquestionable focus of the entire exhibit is actually natural selection, the story of how the environment eliminates some variations and perpetuates other minute variations until they constitute new populations. A plaque next to a series of mouse photos

Figure 5.4. The facade of the Natural History Museum, South Kensington, London. The building was originally opened in 1881.

reads: "Dark fur may provide better camouflage from predatory owls than light fur. The dark mouse will thus be more likely to survive to reproduce than its paler-coloured relatives, and over several generations the proportion of dark mice in the whole population will increase. Because environments differ from place to place and may vary with time, over thousands of generations populations may adapt and change again and again." Most of the exhibit consists of different examples of this same point. At the very end of the exhibit, as one is walking out, there is a little wall plaque that quickly acknowledges that there is more to evolution than just natural selection, mentioning that genetic mutation and ontogenetic development are important, too.

In Paris, where things are quite different, you are likely to leave the museum asking, "Charles who?" The reasons for this are complicated. After the high-profile battles in the early part of the nineteenth century between Cuvier and transformists such as Lamarck and Geoffroy, the French perception of evolution issues centered around two basic themes. The first can be termed, for lack of a better name, "internalism," and the second theme can be discussed under the title "empiricism."

Starting with Lamarck's first major law of transformation, that all living creatures had an inherent trend toward the next highest link in the chain of being, the French perceived evolution as a process carried out by individual organisms—a process initiated within themselves and for themselves. Whether it was Lamarckian acquired-trait transmission or Geoffroy's internal-growth laws, the locus of transformational change and success was in the organism. The environment played a negligible role. Furthermore, the changes inside the creature were seen to be relatively big jumps (sometimes speciation could occur in only a few generations) rather than very minute alterations. Remember that Cuvier's attacks on transformation make this clear, because his view of the organism as a perfectly self-adapted integrated entity (an organic syllogism, to use my term from chapter 4) prevented any transformations on pain of death (disintegration).

After Cuvier's other line of attack, evolution was also frequently perceived in the light of the empiricist/rationalist debate. Cuvier went to great pains to argue that his transformist colleagues were doing speculative metaphysics rather than experimental science. According to Cuvier, people who talked about evolution had moved way beyond the testable world of experience; they had entered the realm of imagination, where no interesting research could legitimately follow. The equation of evolution theory with effete metaphysics was a strong association for the French, lasting into the 1870s.

When Darwin's *Origin of Species* was finally translated into French, in the early 1860s, the translator, following the French tradition, placed the work in the context of metaphysics. The translator, Clémence Royer, was highly supportive of Darwin in her provocative introduction, but her support centered around metaphysical materialism rather than the scientific findings. Royer praised Darwin's ideas as inherently antagonistic to the backward metaphysics of religious thinking, and thereby pitched the ideas into a melodramatic arena that failed to interest most experimental French scientists. One of her purple passages reads:

> The doctrine of Mr. Darwin is the rational revelation of progress, pitting itself in its logical antagonism with the irrational revelation of the fall. These are two principles, two religions in struggle, a thesis and an antithesis of which I defy the German who is most proficient in logical developments to find a synthesis. It is a quite categorial yes and no between which it is necessary to choose, and whoever declares himself for the one is against the other. For myself, the choice is made: I believe in progress.

Despite Royer's accolades, this presentation of Darwin still hindered his ideas by allowing the scientific community to dismiss him as yet another nonscientific metaphysician. Darwin himself lamented the French translation in a letter to Charles Lyell, saying that "the introduction was a complete surprise to me, and I dare say has injured the book in France."

The general mood of empiricism that we saw developing in Hunter and Cuvier became even more pronounced in the middle of the nineteenth century, when a zest for experiment led to great advances in physiology. France in particular, during the middle of the century, had the powerful figures of Louis Pasteur (1822–1895) and Claude Bernard (1813–1878) championing the strictly experimental approach to science. Bernard, echoing Hunter's earlier refusal to pursue unverifiable origins, claimed that it didn't matter one bit whether you were a Darwinian or an antievolutionist. Both positions were equally conjectural, nonexperimental, and uninteresting. And Pasteur summed up this experimentalist agnosticism by first listing some of the "many great questions being discussed these days." He states a few of the famous antinomies: "unity or multiplicity of human races; creation of man many thousands of years or centuries ago; fixity of species or slow transformation of species from one to the other; matter reputed to be eternal rather than created;

the idea of God being useless, etc." Then he summarily weighs in on the heady puzzles by stating that "these are all questions that cannot be solved." And he concludes by saying, "I take a much more humble role in tackling a problem that can be solved experimentally."

This positivist attitude put evolution off the scientific docket in France for some years. The irony is that Darwin himself shared this contempt for speculation and metaphysics, trying hard to distance himself from the purely conjectural evolutionism of people such as Robert Chambers and his own grandfather, Erasmus Darwin. Still, Charles Darwin was perceived by the French as another dreamer, and subsequently for them there was no "Darwinian revolution" as there was in England, Germany, and America.

By the late nineteenth century, however, the advances of evolutionary science in these other countries could not be ignored by the French. Darwinism had indeed shown itself to be a very fruitful research program and a reasonable way to understand an array of phenomena. Among the subjects evolution could speak to was human beings. In France it was the anthropologists who first began to take transformism seriously again, beginning to characterize human cultures according to evolutionary ideas. The fact that evolution came back into vogue via the subject of humans certainly influenced its interpretation. The French were not pleased with the Darwinian view of evolution, wherein each organism is at the mercy of capricious external forces (natural selection).

The historian of science Peter Bowler argues that the French preferred alternative mechanisms to Darwin's natural selection. For the French, Darwinism lacked the optimism of Lamarckian adaptation. Bowler points out that "Lamarckism drew upon the notion of the individual's creative response to environmental challenge in order to promote a more optimistic and purposeful view of adaptation." And, in addition to this objection, they believed (perhaps through a combination of Cuvierism and Catholicism) that Darwinian evolution was too lawless—too chaotic—to be true. So as transformism came back into vogue in France, it did so in terms that were familiar; terms that harked back to the more comfortable old-time metaphors of Lamarck and Geoffroy.

Once you appreciate this story, the evolution exhibit at the Grande Galerie is no longer mysterious. According to the exhibit, evolution is really a French idea, to which Darwin contributed a couple of interesting concepts. For the French, evolution is not primarily about natural selection. The dioramas that cover the history of evolution claim that Darwin basically strolled down a path originally beaten by Buffon, Lamarck, Cuvier, and Geoffroy. Yes, even Cuvier is placed as a forerunner of evolution, and today's French patrons are never presented with the facts of his bitter hostility to evolution. This treatment of Cuvier is just revisionist flummery. But the French are correct in their general conviction that transformism is a much bigger and older idea than Darwin's natural selection. The current exhibit embodies the long-standing view of Darwin as merely continuing a French tradition.

Given France's lukewarm embrace of Darwinism, even to this day, one might wonder how they can have a Grande Galerie de l'évolution. If curators are not filling the exhibit spaces with testaments to natural selection (as the London museum does), then what are they filling them with? The answer, strange though it sounds, is mutations.

It has to be remembered here that these contemporary differences between evolutionary museums are matters of emphasis, not complete exclusion or inclusion of subjects. When I point out, for example, that the AMNH stresses cladistics, I'm not saying that the Field Museum and the South Kensington museum have nothing on the topic. But while the AMNH's entire floor plan is given over to the teaching of cladistics, the Field museum devotes a small series of nested cardboard boxes (clades), each containing rubber toy animals, to the same subject. The Field acknowledges cladistics (it can't afford not to) but does so without making it the centerpiece of the museum's organization. And just as this is a matter of selective emphasis, so too the French museum acknowledges natural selection (even including the classic "industrial" melanic moth scenario) but severely downplays its importance.

The Grande Galerie devotes much of its third level to educational displays of genetics principles and mutation experiments. Many displays feature large photos of normal and abnormal fruit flies and the lab equipment associated with them. By my calculation, less than 10 percent of the total space on that level is devoted to the idea that changing environments modify and regulate species (natural selection), and even this small space (purportedly dedicated to natural selection) is half filled with descriptions of the transmitting gene chemistry. The choice to accentuate mutation may seem strange at first, but it is actually a natural move for the French.

Before the science of mutations was absorbed into the Darwinian paradigm, during the 1920s, mutation was proposed as a viable counter-Darwinian mechanism of evolution. In the early part of the twentieth century some geneticists believed that the new experiments being done on primrose plants and fruit flies revealed non-Darwinian evolution. This experimental approach used everything from salts and acids to radium to influence the birth of mutant flies. Mutants that arose, either naturally or unnaturally, were bred and tracked for their influence on further generations. The Darwinian mechanism of natural selection claimed that variations occurred in a gradual and continuous way, with speciation developing as a result of gradual editing by the external environment. But the early mutationists believed that they had some evidence of larger-scale jumps; evolution might happen by discontinuous leaps rather than very slight variations. Natural selection, on this view, would not disappear altogether, but some early mutationists believed that genetic mutation (not the environment) was the true shaping force of evolutionary change.

The early mutationists explicitly stated that their novel mechanism of speciation undid some of the distasteful implications that they perceived in Darwinism. With natural selection demoted and the idea of limited resources deemphasized, many of them believed that evolution could be seen as a less competitive and brutal process. The mutationist Thomas Hunt Morgan, for example, claimed that his theory gave us a different view of nature: "Such a view gives us a somewhat different picture of the process of evolution from the old idea of a ferocious struggle between the individuals of a species with the survival of the fittest and the annihilation of the less fit. Evolution assumes a more peaceful aspect." Mutationism also gave some consolation to those internalists who felt uncomfortable with the external causation of natural selection. Those who perceived natural selection as "lawless" (clearly a misperception) also approved of mutationism because it seemed to conform to predictable Mendelian ratios of trait distribution. And on top of all this, the new mutation science was experimental in a way that made scenarios about past selection events look speculative by comparison.

But after the initial excitement over mutations, it began to seem more and more like an extension of Darwinism rather than its opponent. The early promise of mutationism was based on deceptive experiments that suggested viable large-scale, one-generational changes. Further research showed that real mutations are usually very small-scale affairs, with alterations in the structure of existing genes occurring much more frequently than the sudden appearance of "hopeful monsters." In the 1920s research showed that most of these rare large-scale mutations were lethal or deleterious. In addition, mutationism could be an alternative to natural selection only if the mutations arose in a nonrandom, directed way; hitting on adaptive traits quickly without the eliminative competition of finite resources. But the more that scientists experimented with mutations, the more their random, nondirected character became obvious. The mutations don't discriminate; nor do they precognize and thereby propose only the "correct" kind of trait change. Selection is still the shaping force that molds the mutations into viable population traits.

For the British and Americans, mutation science became a part of the neo-Darwinian synthesis in the middle of this century. But the French, having never been fully converted to Darwinism, tend to present evolution as a pluralistic collection of alternative mechanisms. Mutation, for the French, is not reducible to, or subsumable within, Darwinism; mutationism still offers some culturally important alternatives for them. According to Peter Bowler, "even today . . . the majority of French biologists refuse to take the modern synthesis of genetics and the selection theory seriously." Of course, there is no explicit discussion of any of this in the museum displays themselves, but once you understand the cultural commitments and the cultural history involved, then even these contemporary evolutionary exhibits are unmasked as cultural microcosms.

CHAPTER 6

EVOLUTION AND THE ROULETTE WHEEL: A CHANCE COSMOS RATTLES SOME BONES

AFTER COMPARING THE GALLERIES IN New York, London, and Paris, it was increasingly clear that evolution itself is constructed differently depending on which museum you happen to be visiting. One of the things that emerged out of this comparative pilgrimage was a heightened sensitivity to which subjects were being included in exhibits and which subjects were missing. In a rather obvious sense, of course, it is hard to see what's missing in an exhibit, because it isn't there. But comparative visits to many museums can actually turn one's eye into a perceptive decoding device. And one of the striking omissions that my salubrious decoding eye noticed was that the French and British evolution exhibits never really mention religion anywhere.

Now, perhaps this omission is right and fitting; after all, we're at the museum to learn about nature, not hear a sermon. But one would need to have been living in a very deep cave to be unaware of the 150 years of constant battling between religion and evolution. The London museum had exactly one line on the subject in its "Origin of Species" exhibit: "Many people do not believe that there is a conflict between evolution and their religion." That's it—no elaboration, no follow-up. And the French Grande Galerie was even less interested in the topic. Even though early French discourse on transformism was quite explicit on the metaphysical dimensions, you would look in vain around today's Grande Galerie for a mention of either religious conflict or conciliation. It seems puzzling, given the Catholicism of France, that there should be no coverage of this prickly topic.

To some extent, of course, the truly vociferous battle between evolution and religion is a homegrown American phenomenon. Christian fundamentalism, the Scopes monkey trial, and the notion of equal time for creation science are all American phenomena; they are both causes and effects of the American psyche. Having grown up in America, I had unconsciously expected to see evolution portrayed in the familiar costumes, namely, the momentous struggle between godless humanism and righteous piety. But even acknowledging the true cultural locus of the melodrama, one must recognize the deeper metaphysical tensions, tensions between chance and design, that have permeated most Western cultures since as far back as the time of Democritus.

Inside the entrance of the evolution exhibit at Chicago's Field Museum, entitled "Life over Time," stands a large wheel-of-fortune device, a kind of vertical roulette wheel. The wheel is painted with images that represent genetic traits, and children are invited to spin the wheel to see which trait would be beneficial for an aquatic animal's evolution onto dry land. There are actually two such wheels, one for plants as well as one for animals. Hanging above them is a bold-lettered sign reading "Wheels of Adaptation." Listed on the animal wheel are a combination of obvious adaptations such as "air-breathing lung" and "muscular fins with sturdy bones," as well as some deleterious ones such as "feathery gills that dry easily in open air." Between the wheels is a game-show-host manikin with a beady-eyed fish head, inviting patrons with the Dirty Harry challenge "Feel lucky, amphibian?" It's a funny exhibit (figs. 6.1, 6.2). I smiled as I studied it, thinking vaguely that it would be nice to meet such good-humored curators. While I was standing there, a prepubescent baseball-capped boy was feverishly spinning the plant wheel with all his strength; totally oblivious to the point of the exhibit. On the sign next to the wheels, it says "Adapt or die!" and "Life is a gamble."

After three rooms of hands-on interactive displays for kids to learn about prokaryotic and eukaryotic cell functions and structures, the exhibit opens into a hallway lined with corkboard and equipped with sheets of blank paper and pencils. These cork slates are called "talk-back boards," and patrons are encouraged to write feedback and tack up their respective complaints, confusions, suggestions, and the like for curators and other patrons to view. Some patrons talk back to earlier patron postings. Perusing the feedback, I noticed a few notes of appreciation for such an integrated and intelligent exhibit. Some of the scrawled notes are cryptographic nightmares, but most are fascinating glimpses into the minds of anonymous museum fans.

Tacked up next to the hastier jottings was a one-page typed letter. In the upper left-hand corner it was addressed to the Visitor Service Department of the Field Museum. This is such a wonderful letter that it must be quoted in its entirety:

> This last February my wife and I enjoyed visiting your museum. I especially enjoyed the display on how the Japanese

Figure 6.1. The "Wheels of Adaptation" display, from the Field Museum's "Life over Time" permanent exhibit. A fish-headed game-show host invites patrons to spin the wheels, asking them, "Feel lucky, amphibian?"

Figure 6.2. A close-up of the "Animal" gambling wheel. The interactive display conveys the blind character of mutations and the environmentally relative nature of adaptation. Patrons spin the device to determine whether they will get a useful or harmful chance mutation.

were treated in the U.S. during WWII. It was educational in an historic sense and helped to elevate one's sense of moral responsibility. However, we were appalled at the lack of moral bearings in your "Life over Time" exhibit. Using sweepstakes, roulette wheels, slot machines, dice and so on to illustrate evolution is very inappropriate, especially since most of your audience is children. Gambling is an adult concept and a vice, by the way. It also is an unobjective interpretation of the universe's ontology. For instance, what appears to the human mind as mere randomness, may from a divine perspective not be random at all. You may argue that this is a biased perspective, ridden with personal values and interpretation. However, seeing the principal cause and government of the universe as a roulette wheel is filled with the same personal subjective views as well as being morally bankrupt. Please, please pick a more objective and socially responsible means to convey your concepts of evolution!

Sincerely, Alan

Even though I do not personally share Alan's sentiments on this issue, I love the letter all the same. It is direct, friendly, thoughtful, and perceptive. And, most of all, it acknowledges that museums can have philosophical subtexts. Alan has recognized that in the very structuring of the exhibits, natural history museums can convey more than just factual information. They can also convey ideological assumptions and ontological (metaphysical) commitments.

Tacked next to Alan's letter were two smaller talk-back sheets. One of them, apparently in support of Alan's letter, asked simply: "If it's all random chance, how can 'right and wrong' exist? It is very disappointing to see life as a crap-shoot" (fig. 6.3). The other note reads: "Alan needs to get a life & I'd bet he has no children!" These notes are great because they're heavily punctuated and underscored, and the lettering flares out in places, embodying the authors' passion. The anti-Alan letter continues: "The 'gambling' objects are manipulatives geared towards interaction. Children love these & can identify & understand concepts much easier when they can 'handle' things. Alan needs to understand that life is a gamble. Whenever he walks out his front door, he's gambling with his life!" The central ideas of these letters are worth further exploration, particularly the one by Alan. But first here is a recap of the major types of religious resistance to Darwinism.

A quick perusal of the Internet dredges up some extremist venom. Many creationist Web sites are saturated with anti-Darwinian rheto-

Figure 6.3. A craps table display from the Field Museum's "Life over Time" permanent exhibit. Patrons are invited to throw dice to simulate the chance character of evolutionary speciation and extinction.

ric. A Web site entitled "The Bible Believer's Research Page," for example, contains the following warning: "The theory of evolution is contrary to the Word of God and is thus a lie. Christians who believe they can compromise their beliefs about creation to accommodate evolution need to seriously consider the havoc Satan has unleashed on humankind as the result of evolutionary thinking." One hopes that this position, while prevalent on the Net, is rare in the general public.

A Gallup poll taken in the early 1990s showed that 47 percent of Americans accept the idea that "God created man pretty much in his present form at one time within the last ten thousand years." For lack of a better term, I'll call this fiat creationism, thereby indicating the let-it-be-done quality of this rapid miracle. To what degree these people also commit to other creation science points is unclear. All of them clearly believe in a very young earth, but some of them also believe in Noah's flood and a one-month-long creation of the Grand Canyon.

There is a great little museum near San Diego, in the Institute for Creation Research, that embodies the creationist ontology. Alan might find it most refreshing. Inside the small museum, where dioramas, plaques, and lighting are professionally crafted, you can find an organizational plan that is structured on creationist principles. The first diorama, labeled "Day 1," is a brooding painting and model construction that depicts the earth's creation. Next is "Day 2," the creation of the atmosphere, complete with placard text of verses from Genesis. "Day 5" is a diorama depicting the creation of "the things that flew and the things that swam" (which makes me think that God must have been an ecology-minded taxonomist, rather than opting for the morphological route). Against evolution, there is a disturbing exhibit that contains artistically placed animal skulls and bones. This scarcely lit manifestation of rot makes the point that death itself (which evolutionists see as a natural feature of organic change) is just a temporary punishment that God meted out after the fall. This cheery fact is derived from a Genesis passage that claims that there were no carnivores before the fall. So fiat creationists clearly have their museum. But Alan seems more sophisticated than this, and one suspects that he would pen a similar complaint to the Institute for Creation Research about its lack of objectivity.

I suppose that my last comment implies that potentially 47 percent of the American public is unsophisticated. And that brings us to the really important aspect of fiat creationism. The relationship between this form of creationism and evolution theory cannot be understood as a battle between evidence claims and equally reasonable interpretations of nature. Evolution theory wins the evidence debate with relative ease (see the writings of Philip Kitcher, Arthur Strahler, Douglas Futuyma, Langdon Gilkey, Michael Ruse, etc.). The real battle is just a series of trenches in the American culture wars. Snobby urban college boys like me are looking down their noses at the hardworking middle-class rural folks who have built their lives on family values and belief in God's plan, and we're calling them unsophisticated bumpkins. Contrariwise, these conservative traditionalists see their rural, close-knit lives under siege by liberal ivory-tower eggheads who have undercut the nuclear family and traditional morality by claiming (via Darwinism) that we're all just animals. Obviously, we won't be clearing this up anytime soon.

The same Gallup poll records that another 40 percent of Americans subscribe to the view "Man has developed over millions of years from less advanced forms of life, but God guided this process, including man's creation." This is quite different from the all-at-once miracle of fiat creationism. Using a modification of the phrase first introduced by John Dewey, this version can be referred to as "installment creationism." God's plan is installed in nature over stretches of evolutionary time. There is transformation, but it is channeled by the designing mind of the Deity. Installment creationism can be very sophisticated and thoughtful. The general position contends that while the immediate processes of nature may be purposeless (genetic variations in an animal's gametes, for example, are nondirected and random), it is still possible that the whole law of nature (the big picture of organic evolution) could have some undetected purpose. Now we're getting closer to the thinking behind Alan's letter.

In addition to the belief that God works within nature by a sort of installment plan, many other people have returned to a kind of deism. Like the Newtonian-era God, this Deity is a first cause of natural laws (in our era, the first cause of the big bang) rather than the micromanager of specific temporal events. This first-cause argument usually blends together in people's minds with the more traditional design argument (a designing God can be inferred from the adaptations we see in nature), and the result is a belief that God's actual role in evolution consists in his seeing all the future permutations of matter and then flicking the appropriate first domino of causality.

I want to submit that these different creation scenarios have a very similar rational, emotional, and cultural center. At this center, two interrelated themes emerge. The first theme is that ontological order and structure (antichaos) are necessary. The second theme is that ontological reductionism is bad. Let's examine these two points in turn.

Alan's letter is primarily concerned with the debate between random versus nonrandom ontology. We have encountered this issue before, back when we were examining Cuvier's pursuit of nature's laws. Cuvier's antitransformist ontology was not a mere pledge to religious orthodoxy. It was, among other things, an epistemological issue. If nature were not ultimately orderly and fixed, then we mortals could not have a science of it. Science is a system of rational universal laws, and we cannot achieve that end if nature is chaotically dancing all over the place. As an analogy, you cannot draw someone's portrait if the person will not sit still. In our own century, Einstein (a Spinozist) actually echoes a similar dislike of ontological randomness when he claims that every scientist (qua scientist) must have "faith in the possibility that the regulations valid for the world of existence are rational, that is, comprehensible to reason."

For Cuvier and Einstein, some ontological fixity is necessary as a presupposition to doing science. Despite the obvious dissimilarities between these two titans, they would agree that the laws of nature (which are a summary of causality itself) provide the requisite solid foundations for erecting good science. So too, many moral theorists, including Alan, have extended a similar argument to the ethical realm. Order and nonrandomness, they argue, are absolute presuppositions for morality and meaning. Without the bedrock of a divine plan (containing an objective, unambiguous right and wrong), morality would be a purely subjective affair.

There is a certain epistemic logic to this position. If you want to resolve some factual dispute, such as "Is there or is there not a fly in that soup?" you and your friend take your subjective difference of opinion to the bar of objective reality and resolve the disagreement in a straightforward way (you examine the soup). The idea that there is a divine plan in nature means that moral disputes can be resolved in a logically similar way. That is, if you want to resolve some moral disagreement, such as "Is it wrong for the president to sleep around?" you and your friend try to take your subjective opinions to the bar of objective reality (examine God's plan). But of course it doesn't quite work this way. There is a serious difficulty with how one examines God's plan, which is that you won't find it in a soup bowl or any other aspect of the material cosmos. You'll be hard pressed to find the unambiguous plan in a book as well, because—besides the very serious problem of which book should be followed—it is a plain fact that every moral injunction you do find is a Rorschach blot for interpretation. Observing God's plan does not work in the same way that observing flies in soup does, but one can easily appreciate why we would want it to work.

The scientific dislike of randomness (embodied in Cuvier and Einstein) is quite different, however, from the moral dislike of randomness. For many of the scientists who fought randomness, orderliness just meant that events and things have causal histories; things do not pop into existence, and events are not uncaused. Unlike this dedication to the idea that everything happens by a cause, Alan's notion of orderliness seems to claim that everything happens for a reason. These are quite different claims, though they are frequently confounded. As is clear from our survey of Darwin's theory in chapter 5, the randomness of variation is chaotic only in the sense that mutations are blind to the specific environmental pressures awaiting them. But they are not chaotic in the sense of being uncaused, inherently unpredictable, or unknowable. In other words, the sort of ontological order (causal patterns) needed for science to work is completely present in Darwinian nature.

But Alan, as well as other proponents of sophisticated creationism, is interested in the kind of order that comes not from antecedent causal trajectories, but from a plan. Order in itself is significant for this group only if it is the product of design. And while we have touched briefly on the scientific interest in order and the moral interest in order, we would be remiss if we didn't at least acknowledge the psychological interest. Belief in a cosmic purpose (Alan's clear preference) is a significant balm. Many people—nay, most people—find the idea that we are nothing but atoms in a void rather depressing.

This leads to the second theme that anchors all three creation scenarios. Many anti-Darwinians oppose the theory of evolution on the grounds that it reduces the noble and free human being to the level of the base and instinctual animal. They take an antireductionist stance, maintaining that there is a significant qualitative gap between humans and beasts. Fear of reductionism is a very old phenomenon, and it certainly did not originate with the objections to Darwin. But this age-old fear (Plato argued against Democritus's version) is usually concerned with the same thing, namely, freedom. If we are just material creatures, without spiritual infusion, are we free?

Literature has built a cottage industry on the fears of scientific reductionism. Perhaps Mary Shelley's novel *Frankenstein* is the most obvious case. Here the following point is tacitly raised: If a humanoid creature can be created with inanimate material parts jolted with electricity, then perhaps we too are reducible to such humble physical stuff. Currently, genetic technology is raising this concern all over again. The very heterogeneous attempts to claim that life is more than mere matter are classified as forms of vitalism. But in the late nineteenth century the fears about reductionism focused more around the human mind itself rather than life. The experimental physiologist Claude Bernard, for example, argued strongly for an application of deterministic physico-chemical explanations for all phenomena, even those phenomena that previously fell under the aegis of the supernatural. Near the end of Dostoyevsky's *Brothers Karamazov*, Dmitri, who is awaiting trial for murdering his father, begins to use the name Bernard as a euphemism for "reductionist idiot." He starts to refer to his lawyer as "Bernard" because the counselor is trying to reduce Dmitri's murderous actions either to a bad environment or a physiological brain disorder (and thus not free will). The general sentiment of antireductionism is quite familiar.

Many opponents of Darwinian evolution echo this ancient objection to reductionism. The slope of the belief runs like this: If humans are animals, then we are no different fundamentally from squirrels and pond scum, which means we are just temporary aggregates of inanimate stuff, which means that we are all just atoms in the void. I sharpened the grade of the slippery slope for dramatic effect, but the picture is clear.

There are actually very good reasons to fight against a scientific reductionism that treats every other level of nature as explicable by the microscopic level. That reductionist approach has served us relatively well in the past (since the 1600s), but we should not get carried away. Examining everything in terms of the mechanics of its parts is just narrow-minded science, and many good scientists (predominantly biologists) have called for pluralistic and holistic approaches to the study of phenomena. But apart from this in-house scientific critique of reductionism, we must again acknowledge the psychological issues at stake for the nonscientific version. Just as a cosmic plan is quite bracing for many, the rejection of reductionism is a soothing balm of personal freedom and dignity. The paradoxical relation between the two themes is interesting: A deterministic, orderly plan must underlie the cosmos, but not the human will.

Only a few yards from Alan's letter at the Field Museum is a glass display case devoted to newspaper clippings, plaques, and photos. Tacked to the display board inside the case are sections from the *Chicago Tribune* and *Chicago Sun-Times*, as well as the *New York Times* and others. There is a photo of John Paul II waving a blessing to some throng. Headlines read "Pope Backs Darwin—Within God's Plan," "Pope Bolsters Church's Support for Scientific View of Evolution," and "For Pope, Faith and Science Coexist."

In the late 1990s, a few years after he apologized to Galileo, Pope John Paul II delivered a written message to the Pontifical Academy of Sciences stating that the Catholic Church formally recognizes the truth of evolution theory. The Pope emphasized that, due to the weight of scientific evidence, the idea of evolution could no longer be only provisionally accepted as a mere hypothesis. Evolution, according to John Paul II, is a fact. The Pope's declaration was further clarified by Monsignor Francis Maniscalco, a spokesperson for the National Conference of Catholic Bishops, who said, "What belongs to science belongs to science, and what belongs to religion belongs to religion." And by this tautology, he seems to mean that the claims of evolution and the claims of Catholicism do not trespass on each other's territory. The findings in one domain should not, and cannot, controvert the other. In other words, Catholics do not need to fear evolution, and evolutionists do not need to fear Catholics.

In 1950 the Cold War Pope, Pius XII, paved the way for John Paul's pronouncement by claiming (in his encyclical *Humani generis*) that evolution was not incompatible with Christianity but that the flock should remain on the lookout because atheists and Communists seemed to like the evolution idea, too. Generally speaking, Catholics have never had much animosity toward Darwinism because they read the Bible, particularly Genesis, figuratively and symbolically rather than literally. In 1981, for example, John Paul II said that the Bible "does not wish to teach how heaven was made but how one goes to heaven." Only a relatively small group of Protestants, distinguishing themselves as fundamentalist, see the Bible as a work of journalism. And naturally, fundamentalists were not amused by the Pope's recent clarifications. Bill Hoesch, a spokesman for the Institute for Creation Research (where they house the creationist museum), responded by saying, "[T]he Pope has a lot of followers, and if he said the moon was made out of green cheese, a lot of people would believe him on that too, but it doesn't mean it's true." One can almost hear in Hoesch's comment the bitter lament of someone losing a putative ally.

The Pope has put his imprimatur on evolution theory, but not without an important caveat. This goes back to the point about antireductionism. The Pope still claims that evolution does not give enough weight to a person's spirit and that by itself, evolution is "incapable of establishing the dignity of man." Evolution is true, but the human soul is one of those godly installments that I mentioned when discussing installment creationism.

Several months after the Pope's media blitz, the preeminent American Darwinian, Professor Stephen Jay Gould, wrote an influential article in *Natural History* magazine entitled "Nonoverlapping Magisteria." The article was subtitled "Science and Religion Are Not in Conflict, for Their Teachings Occupy Distinctly Different Domains." In the piece, Gould mulls over the Pope's recent statements and uses them as a catalyst for a general theory about the relationship between science and religion. Using a Latin term that he adopts from Pope Pius's *Humani generis*, Gould describes the different domains of science and religion as different "magisteria." A magisterium, which comes from the word for "teacher," is a domain of teaching authority. For example, one might argue that the American separation of church and state is itself a division of magisteria. Gould uses this concept of magisteria to explain that science and religion can never really come into conflict. The idea seems to be that two groups can be antagonistic toward each other only if they have some common thing at stake; remove that common thing, and the clashes become ephemera. Gould supports the Pope's demarcation of two totally separate realms, evolution and religion. He explains:

> No . . . conflict should exist because each subject has a legitimate magisterium, or domain of teaching authority—and these magisteria do not overlap (the principle that I would like to designate as NOMA, or "nonoverlapping magisteria"). The net of

> science covers the empirical universe: what is it made of (fact) and why does it work this way (theory). The net of religion extends over questions of moral meaning and value.

This handy acronym, NOMA, nicely encapsulates an attitude of toleration and cooperation. In fact, besides describing the professional relations between clergy and scientist, it neatly characterizes the way that these two domains can seem to coexist in the same person. Many people are able to compartmentalize their own commitments into areas that do not infringe on each other. The overwhelming majority of scientists, for example, consider themselves religious people as well.

Gould seems to resonate with the phenomenalist tradition that we saw emerge with John Hunter and Georges Cuvier. Remember that after the Enlightenment's characterization of nature as a seamless system of deterministic laws, there wasn't any room for things such as miracles and souls. The borders of religion were redrawn, and instead of its old domain (wherein it could make pronouncements about the material cosmos), it was reassigned to the realm of first-cause theology. Religion was encouraged to steer clear of the proximate material cosmos, and after Darwin's natural selection finished off the design argument, religion was completely reassigned to the morality domain. Like the others in this two-realm tradition, Gould doesn't think you can learn anything about morality from studying the material world, nor can you learn anything about the material world by studying religion and morality. This seems reasonable. It does not seem likely that you can figure out what your duties are to your spouse by doing science, and you cannot figure out how the electrovalent bonding of ions works by looking through the Bible.

Professor Gould would never compromise evolution education by suggesting that creationism should have equal time in the classroom, but he is comfortable with the idea that design ontology ultimately can be compatible with biology. The first case would be a crossing of the NOMA line, but if someone wants to believe that God designed the big bang because it makes her feel better to believe so, then Gould would not object. Nothing scientific is at stake in such a case, and no conflict can arise. For Gould, moral positions are not derived from the empirical universe, so no findings in the empirical universe can contradict or support them. In discussing the Pope's declaration that humans still possess a supernatural soul, for example, Gould states that no scientific research can weigh in on any side of this topic. Souls are entities for the religious magisterium, and the scientific magisterium (the empirical world) has exactly nothing to say on the matter.

The message of NOMA is so pleasant that it is only with great regret that I now criticize it. Who wouldn't want such a harmonious compromise? Religion and science, on this view, could be cordial neighbors. After all, people who are deeply religious do not want to deny the obvious truths of science, and they do not want to be perceived (or perceive themselves) as backward. Likewise, devotees of science do not want to be intolerant killjoys who back people into corners at cocktail parties with hope-crushing arguments, only to mutter darkly to themselves after being abandoned.

Gould admits near the end of his article to a private suspicion that "papal insistence on divine infusion of the soul represents a sop to our fears, a device for maintaining a belief in human superiority within an evolutionary world offering no privileged position to any creature." I happen to agree with this assessment, but I have to call attention to some confusions.

Apart from the fact that Alan, the Field letter writer, would object to Gould's assessment (on the grounds that a "world offering no privileged position to any creature" sounds awfully ontological), it is legitimate to ask Gould exactly how he comes to hold this private suspicion about the soul. He does not directly address this issue. He claims that the whole subject of the soul falls outside the magisterium of science. So that means, by his own definition, that pronouncements on the soul (including his own) are not derived from the realm of the empirical. But if they do not come from the world of experience, then where *are* they derived from?

Gould does not explain the source of his ontological commitments; he just identifies the domain from which they do not originate. Let us remove Gould from the picture for a moment and simply reflect on the ontological commitment, a commitment that claims many adherents. How does one come to believe that the soul does not exist? Here, stated in no particular order, is a nonexhaustive list of pathways that might lead to such a conviction. (1) One might develop suspicions if one cannot detect such an entity with any of one's senses. (2) One might study intellectual history (Eastern and Western religions and philosophies) and come to understand the origins of and uses for the soul concept, none of which may seem applicable in one's own era. (3) One might study enough about science and brain physiology to see that a soul concept is not necessary to explain things such as personal identity, dispositions, emotions, and so on. (4) One might study psychology to appreciate why the soul concept is such an attractive idea—why one might want it to be true. (5) One might analyze enough of one's own behavior, the behavior of others, and commonsense logic to know that nothing ontological follows from wanting the soul to exist, except wanting the soul to exist. (6) One might reflect on one's experience of caveat emptor and notice that many people who strenuously push the "soul idea" have something unhallowed to gain from doing so. (7) One's experiences with the power of the human ego might lead one to agree with the Buddha's claim that belief in an immortal soul actually leads away from, rather than toward, moral action, because it distracts people with more self-centered metaphysics.

The list could go on, but there is no need. It may not be any single experience that leads one to this ontological conviction; rather, it may be a cumulative phenomenon. The issue that I hope this list illustrates is that it is in fact the empirical world from which one derives one's ontology. One may not be employing microscopes and telescopes and bar graphs in this study, but it is experiential nonetheless. One studies causes and effects and consequences, makes predictions, looks for consistency, consults peers, follows through theoretical and practical implications, and so on. It all focuses on human beings instead of geological strata or chemicals (i.e., humans are the subjects), but unless one is willing to claim that human subjects are not a part of nature, then it's clear that one can derive a conviction about the absence of souls from an empirical (taken in its precise sense) study. And like other empirically based convictions, this one is fallible and provisional.

The point here is not to convince anyone of the correctness or soundness of this ontological viewpoint on the soul. One should notice the form of the argument instead of the particular conclusion. What's important here is that, contra Gould's statement, I think that people *do* derive their religious convictions (or irreligious ones, as the case may be) from experience. The religious ideas are always extrapolations from direct experience—extrapolations that are then turned over to our imaginations for further articulation and manipulation. The point is that they are not Platonic innate ideas or a priori entities. It may be presumptuous to suggest this, but I think that it's through encounters with the experienced world that Gould himself forms his ontological convictions. And I say this with some confidence, because I can't even guess where else he might have gotten them.

It seems to me that the same can be said about the other subject that Gould places in the nonempirical magisteria, namely, morality. Gould wants to avoid the classic problem, so lamentably played out in recent history, in which people justify their pernicious, unethical actions by pointing to the brutality of nature. In order to avoid the evils of social Darwinism and sociobiology (which have looked to nature for guidance on moral issues), Gould just ropes off the entire magisterium of experience. I think this response is too draconian. Our moral intuitions, tendencies, and transformations surely arise out of, or are influenced by, our experiences with facts about the world. This includes theories about, and observations of, human nature, encounters with cultural traditions, and so forth.

In his article, Gould acknowledges that some people commit to a cosmos that contains goodness as part of its deep structure. He then distinguishes his own position.

> As a moral position (and therefore not as a deduction from my knowledge of nature's factuality), I prefer the "cold bath" theory that nature can be truly "cruel" and "indifferent" . . . because nature was not constructed as our eventual abode, didn't know we were coming (we are, after all, interlopers of the latest geological microsecond), and doesn't give a damn about us (speaking metaphorically). I regard such a position as liberating, not depressing.

Once again I find myself in agreement with Gould's worldview here. But while I know how *I* arrived at it, I'm not sure how he did. Does Gould really want to contend that his own empirical research, as a scientist and as a human being, has had *no* impact on the formation of his moral position? Isn't it his own investigations into evolution that have given him the cold-bath theory? If not, where did he get it from? Was he born with it? Did it magically appear in his mind, without any observations or sensory inputs from the empirical magisterium? If Gould's moral and ontological commitments are not founded on any experience of nature and the beings that populate it (including humans and their history), then on what grounds does he commit to the cold-bath view and not some other (or, indeed, the very opposite) view?

Strangely enough, Alan the letter writer and Gould have some things in common. Alan complains about the view of life as random or a gamble, offering his own palliative ontology as an alternative. But then he catches himself and claims that both he *and* the Field Museum could be correctly accused of importing subjective biases into the discussion of nature. He concludes by asking the Field to return to more objective representation techniques. Gould's NOMA idea seems to agree with Alan's view that there is an objective database of experiential facts, a database that is completely independent of ontological and moral theories. Therefore, strictly speaking, virtually *any* metaphysical and moral viewpoint can legitimately coexist with that body of empirical information, because the one has no overlap with the other. Because scientific data never touch on ontology (either to controvert it or to buttress it), any ontological system is equally subjective and tenable. Both Alan and Gould seem to accept a strict fact/value distinction. Since facts are derived from our experience of the world, then the remaining province for values (such as cold-bath ontology versus warm-and-fuzzy ontology) is something like personal taste. I have deep suspicions about the cleanliness of this fact/value distinction.

The NOMA proponent might acquiesce, on reflection, and agree that some experience with other people (their motives, actions, consequences) has led him to inductively formulate some convictions and rules of conduct. But if this is what we are doing when we formulate our moral and ontological creeds (and I think it is), then why isn't this just empiricism? Epistemologically speaking, it is derived from the empirical universe. Perhaps the NOMA proponent will counter that empirically based ethics and ontology are concerned with a different *kind* of experience; one has to do with analyzing fossil strata and the like, and the other has to do with analyzing the consequences of betrayal and so on. These are different, yes, but not as different as the NOMA proponents think. The ethical business of trying to understand, respond to, avoid, or prevent betrayal is not wholly independent of the world of facts. People might quarrel about where to find the most relevant facts (literature, anthropology, psychology, etc.), but we should not think that this value area is completely nonempirical.

To my mind, there are not two different kinds of experience—not a realm of scientific experience and a separate realm of religious/moral experience. There is just experience. Period. What inferences are drawn using that raw stuff (experience) and what interpretative strategies are employed can definitely differ enormously. But the process of analyzing experiential events or testing consequences of actions and learning from them (basing provisional commitments on them) is not incommensurably different in science and in ethics.

Ontological and moral commitments are not *just* biases, though they *can* be biases. Unlike matters of taste, the value perspectives that Alan raises in his letter can be legitimately assessed, weighed, and judged. There are, for example, two arguments that the Field Museum can use to defend itself against the charge of biased exhibition. Alan's letter expresses a very common objection to a chance-based cosmos when he says, "[Gambling] is an unobjective interpretation of the universe's ontology. For instance, what appears to the human mind as mere randomness, may from a divine perspective not be random at all."

There is a significant problem with this argument. Logical possibility is *not* equal to empirical observation. The former cannot be used to negate the latter; it can only be used to keep our minds open. Alan thinks that it is logically possible that the cosmos is divinely ordered, and he's right. It is logically possible that the world is divinely ordered; and it is also logically possible that the world is infinitely more evil and corrupt than the human perspective can know, and it is also logically possible that Elvis Presley was the second coming of Jesus Christ. Granting the one logical possibility is tantamount to granting the others, because logical possibility simply means that there is no internal contradiction in your proposition. Having some hypothesis turn out to be logically possible is just the beginning of an interesting idea; next, you have to do the empirical work to see if the idea can withstand the gauntlet of experience and remain compelling. Nothing interesting follows from mere possibility, and the fact that something is possible is not a sturdy peg on which to hang your ontological hat—everybody else's hat is already there. Alan and similarly minded design proponents are suggesting that we should take our actual human perspective (in which we observe random variation in the evolutionary process) and weigh that equally with mere possibilities (the possibility that it is not random). But it would be very irresponsible, not to mention impractical, for a natural history museum to devote exhibit space to merely logically possible scenarios.

There is another closely related argument that allows the Field to defend the rhetoric of its roulette wheel. Alan says (and there are many who agree) that the human perspective sees things as random but that the Deity might see them as nonrandom. Now, if you admit that humans can perceive things only from their human perspective (as Alan's letter suggests), then shouldn't our belief commitments follow those perceptions? The Field Museum's presentation (or commitment) to a random nature is based on our scientific encounter with the biosphere—it is based on human perception (because it cannot be based on anything else's perception). The Field's commitment to randomness is superior to the commitment to design, because the latter commitment is actually in violation of the human experience of nature. If it is not possible for humans to see with nonhuman eyes (and surely it is not), then why pretend otherwise and then base convictions on it? The Field Museum offers a view of reality as random because that is what its curators actually observe with trained human perceptions. Aligning your beliefs to your actual and possible experiences is called being reasonable. Aligning your beliefs to something that *cannot* be experienced is bizarre and much less socially responsible than roulette wheels.

The American philosopher William James, in his essay "The Will to Believe," offered up a relevant rule of thumb. Current proponents of NOMA seem to draw upon the spirit of James's famous live-and-let-live philosophy. James argued that an ontological commitment, such as belief in God or belief that the universe has a hidden rational purpose, is justified if it makes some positive difference in the believer's life (e.g., inspiration). This is a pragmatic approach to metaphysical disputes.

This position seems reasonable and charitable, but even James makes the all-important caveat that the pragmatic justification applies only if the evidence for and against the belief is equal. This is a rational constraint on the potential carte blanche of our personal useful fictions. How does this apply to Alan's belief in hidden purpose? Alan would be justified in maintaining his belief *if* the evidence for both his and the Field Museum's position was equally compelling. I have been suggesting that it is not. There is no evidence in biology for a hidden-purpose thesis; there is tremendous evidence that populations evolve as a consequence of nonpurposive mechanisms (random mutation and natural selection). Because of this, I've submitted that Gould's cold-bath thesis and the Field's message of randomness are, in fact, superior or preferable to Alan's chosen commitment in the sense that the first two accord best with the evidence.

As has been argued many times previously, it is very difficult to see how one can reconcile the idea of the traditional purposeful Deity with the brutality and the wastefulness of natural selection. It might make sense to say that God made the laws of physics; he composed and installed $E = mc^2$, thereby giving the cosmos an orderly and elegant logic. But it's quite another story to say that God made the law of natural selection, whereby many more organisms are born than can survive, and the majority of those organisms (which happen to be relatively ill equipped for a specific environment) must incessantly die off in an often painful manner. Nature does not work to minimize suffering and maximize happiness. Should we thank the *E. coli* bacteria in our guts that help us to digest certain foods? Should we, alternatively, blame the virus that is breaking down our immune systems and spreading through the host population? These organisms are not evil or noble creatures, intentionally wreaking havoc or health; they are simply doing what comes naturally, which is to say, reproducing. This is not meant to sound callous or insensitive, for it is obvious that our struggle with other organisms matters a great deal to us, causing real delight and real despair. But from the more general evolutionary perspective, this drama is value-neutral. Suffering, for example, is an unintended but automatic outcome of a predator's natural survival and reproductive techniques.

Darwin himself originally drew attention to disturbing cases such as the ichneumon wasp. This wasp lays its eggs inside the body of a living caterpillar, and when the eggs hatch, they eat the still-living caterpillar from the inside out. The contemporary zoologist Richard Dawkins refers to the Ichneumonidae and points out that the recipe for life is to maximize DNA survival, and happiness is not included anywhere in that recipe.

> So long as DNA is passed on, it does not matter who or what gets hurt in the process. It is better for the genes of the ichneumon wasp that the caterpillar should be alive, and therefore fresh, when it is eaten, no matter what the cost in suffering. Genes don't care about suffering, because they don't care about anything. If Nature were kind, she would at least make the minor concession of anesthetizing caterpillars before they are eaten alive from within. But Nature is neither kind nor unkind. She is neither against suffering nor for it. Nature is not interested one way or another in suffering, unless it affects the survival of DNA.

It is hard to reconcile these ugly facts with Alan's idea of a divine purpose. But perhaps this is where the NOMA position is more nuanced. Perhaps what Gould, the Pope, and other NOMA proponents are saying is this: If you can live with the seeming contradiction between your perceptions of a wasteful and purposeless nature and your belief in a hidden order, then more power to you. *I* don't know how you do it, *you* may not know how you do it, but you do it. In other words, does being rationally consistent really matter in *this* philosophical realm? Perhaps the assumption here is that Alan's belief cannot hurt him in any significant sense (it's only metaphysics, after all). But perhaps having that ontological commitment taken away would be highly damaging. A little reflection, however, reveals how odd this is. Is it really possible to have a belief that could only help but never harm its possessor? Actually, there *is* a phenomenon, or a family of phenomena, that seems to possess this strange character, namely, art.

We usually think that appreciation of a particular kind of music (or art in general) is either neutral or beneficial, and that damage occurs only if the possibility for that appreciation is removed altogether (that is, if the music is censored indefinitely or otherwise denied to us). Our lives are only enriched by our taste in music, and, following the old Roman motto "De gustibus non est disputandum," we don't think that polka fans would be "improved" if they could be rid of their "confusion." It is quite possible that, like religion, the polka fan's taste is a strictly private matter; it is harmless when asserted, but its removal (through antipolka arguments and campaigns) would only diminish the person. I suspect that this is Gould's ultimate point in his invocation of the NOMA principle. Science is public and religion is private. Like musical taste, religion is an intensely personal business; and when one realizes this truth, many science-religion disputes fall away.

It seems reasonable and charitable to assent to this spirit of the NOMA movement, but now one has to recognize certain further difficulties. First of all, most religious people feel strongly, despite Gould's characterization, that they are doing much more than just "morality." Rightly or wrongly, they perceive themselves as offering more than a list of dos and don'ts, though this is always integral. They see themselves as offering the truth about important things such as history, origins, destinies, and so on. And once they begin in these areas, which is immediately, they are quickly out of the realm of the private and into the public. Second, religious people (the Pope is a case in point, but pick anybody, really) are very far from reconceptualizing their religious views on the model of personal artistic taste. Unlike musical disagreements, religious disagreements still cause tremendous bloodshed. While it's true that people who considered their religious convictions to be personal expressions of aesthetic inspiration would be less likely to kill or alienate people who disagree with them, it is also true that such a cultural revolution does not appear anywhere on the horizon.

Close to the end of the Field Museum's "Life over Time" exhibit, over another talk-back board, there is a plaque that asks in bold letters: "Who are we? Just animals, or something more?" Contrary to the French and British museums, the American museum practically baits you into getting your antireductionist dander up. In truth, of course, the Field is not being unduly provocative; it is only honestly acknowledging a cultural anxiety that runs deep for many Americans. One suspects that these issues run deep for those European cultures as well, but the play of these tensions does not occur on the stage of the natural history museum. In smaller type on this Field Museum plaque the text continues by making a very familiar distinction: "Science asks and answers the questions 'How?' How did modern humans come to be? Evidence for the scientific answer is overwhelming—we evolved from earlier not-quite-human relatives. Science does not answer the question 'Why?' That's for other kinds of thinkers to answer."

This how-versus-why distinction is very popular. It functions like a pressure-release valve that allows disagreeing factions to ward off opponents who want to usurp territory. It is a face-saving move that allows scientists, on one hand, and theologians and philosophers, on the other, to have their respective autonomy. It also helps precocious college students get along with their wise, sermonizing grandparents (in other words, the distinction may be a palliative in the culture wars as well). But while it is a common distinction, it is not clear that it is a good one.

I once attended a conference sponsored by the National Science Foundation for college biology teachers. I was the only philosophy professor in attendance, and caused considerable stir by just showing up. One of the things that the biologists kept saying to me over and over again, usually during late-night drinking bouts, was that they were answering "how" questions, but I was answering "why" questions. This was always said with genuine charity and good cheer, so I never had the heart to tell them that I didn't really understand what they were talking about. The main function of a platitude is to close off dialogue, not open it. While I asked them many questions about how this or that physiological process occurred or how this or that animal behaved, they never once asked me why we're all here or what it all means. I'm glad, because I had no answers to give in response to such queries. I also did not want to be the turncoat who exposed the shortcomings of my own brethren by having to tell the biologists that philosophers have no "why" answers at all. But of course, the biologists already knew this in some sense; that's why it stayed off the conversational agenda.

People should not be looking to philosophers for the answers; we're the annoying gadflies who keep asking the questions. Philosophers perform a vital function because our job is to get people to stand back from and reflect upon the human condition. They clarify concepts and construct meaning, but while "why" questions and answers certainly arise in such activities, they don't do so with greater frequency or urgency than in the sciences. The answer, for example, to a "why" question such as "Why is my skin this color?" is natural selection; the honest answer to the question "Why is there natural selection?" is a very long silence. And of course some scholars and some theologians will jump in at precisely this juncture with some comments, but that's not because they have an answer—it's because they can't stand the silence. The distinction made on the plaque, or at my NSF conference, or in the wider culture is really more like a concession to the amateur philosopher inside us all. For Americans (not Europeans), the word *philosopher* has come to mean the opinionated homunculus in each individual's cranium. In other words, the how/why distinction allows the person who is making the distinction to preserve whatever ontological viewpoint her inner philosopher wants to hold, and it can be preserved immune from evidence of any kind. The how/why distinction is not a defense of philosophy or a defense of science; it is a disguised defense of individualism and, in some cases, unfortunately, of dogmatism.

Throughout my exploration of museology, I repeatedly came across hidden agendas at work in the curatorial process (the invisibles within the visibles). And in his letter Alan had raised some compelling questions about what is objective and what is bias in evolutionary displays. I decided to track down the specific curator of the Field Museum's "Life over Time" exhibit and see whether or not he could decode some of this subtext for me. With a little research I was able to contact Dr. Eric Gyllenhaal, and we arranged a meeting one afternoon.

Eric is a forty-something, stout, short man. He wears wire-frame glasses. He's bald on top but sports a ponytail. He has a long gray-brown beard and a walrus-type mustache. He has an enthusiastic demeanor and speaks with a compelling blend of excitement and precision. We met, amidst multiple school groups, in the Field Museum's noisy cafeteria and settled down to sort things out.

I explained to Eric that I had been studying his "Life over Time" exhibit and that I was interested in whether or not curatorial biases had crept into the rhetoric of display. I wanted to eventually ask Eric about Alan's letter—to get the actual recipient's responses to Alan's charges—but I thought I'd start small, with general questions about bias. I had a series of notes from my walk-through, and I started by asking about the titanotheres exhibit.

Titanotheres are gigantic creatures that look something like rhinos. They are now extinct but lived during the Eocene and Oligocene epochs (fig. 6.4). Their name literally means "gigantic beast," and the appellation is fitting. The Field has reconstructed a family of these creatures based on skeletal remains and displays them in the second half of the exhibit. The artistic depiction of these creatures is stunning; they look very lifelike. Sleeping at the bottom of the display of figures is a cute "pup" titanothere, the size of a full-grown human. The pup is nuzzling up to the giant mother, who is also quietly reposing on her belly; the mother's eyes are rendered in a kindly and soft way, and her ears are downturned and relaxed. Standing over them both is the enormous male titanothere, the size of eight men; the beast is rendered in a fierce protective posture, with angry facial features, erect ears, and tense body. A small plaque near this nuclear family contains the following text: "Mammals make good mothers: All mammal mothers feed their babies milk, and care for them as they grow. Many mammal fathers share child-rearing duties—because of that, we showed these titanotheres in a family group. But lots of mammal fathers don't even wait around to see their babies born." This last line got me to thinking. How do they know whether or not these extinct animals had nuclear families? Is the display biased in favor of the nuclear family?

I started to describe this display to Eric, and he interjected, "Mama, papa, and baby, right?" He laughed and pushed his glasses up. "Yeah, I know." He was quiet for a moment, shaking his head.

I prodded him. "Is that a bias, or am I overinterpreting things?"

"Oh, sure. That's a bias. You'll see that when you look at the grizzly bear exhibits, too" (fig. 6.5). "That's something that crept into all kinds of exhibits at the Field Museum back then. This whole business about 'family values' goes back a lot longer than we thought. And really, the people who were doing the exhibits should have known better, and they may have known better and still done it for some good reason, but I've never seen it investigated. But if you look at the bear exhibits, there's always the mom, the dad, and the babies, and in real life the mom would drive the dad away, because he would kill the babies. The titanotheres may have had a similarly rocky family structure; we're not sure."

Eric continued, "We wanted to include the titanotheres in the exhibit in the first place because, well, they were so cool. So that's an example where that part was driven by the fact that we had these incredible specimens, and then we wanted to say something in the label that was short enough that people would read it, and still make some interesting comments that would relate back to the fact that we're in an epoch, at this juncture of the exhibit, where mammals are starting to diversify and all that kind of stuff. So I wrote that plaque label about the mammal moms. And, actually, the first version of the text was a little bit stronger, with reference to mammal dads not sticking around, but the particular scientist who was consulting on that exhibit said it was too strong. He thought we should downplay it a little bit, because I guess we insulted his 'family values.'" Eric chuckled, adding, "Are we encouraging fathers to leave their children by writing this plaque?"

"So that case," I offered, "was a conscious decision to convey a subtext about the naturalness of the nuclear family."

Figure 6.4. Profiles of the titanothere family (Eocene and Oligocene epochs) from the Field Museum's "Life over Time" permanent exhibit. A speculative nuclear family structure is presented.

Figure 6.5. The mounted grizzly bear family from the Field Museum's mammal exhibit. A completely bogus nuclear family structure is presented.

"Yeah, there are a lot of conscious biases in the exhibits. Actually, we call them 'messages,' but they're conscious biases. And then there's always unconscious things, and it would be exciting to see people point out something we didn't know about, because there was a lot we did that we *did* know about. There are some display decisions, for example, that end up having outcomes which we never anticipated; and I don't know whether we do that consciously or unconsciously, but sometimes they just happen because the planning process is so complex. And then, you know, there are just things that we do for one reason that wind up having an effect that we didn't really plan. The talk-back boards are a great way to see how the public is reading some of this hidden stuff."

"Well, on this same issue, then," I asked, "does the overall spatial layout of 'Life over Time' have some conscious or unconscious message? I mean, the exhibit seems to be constructed in a giant **U** form, and the largest room was the dinosaur room, which was located between the other wings."

"That's actually slightly smaller than the other two rooms," Eric countered. "Yeah, it may not seem like it, but the dino room is about six thousand square feet, and the other two galleries are seven thousand square feet."

"Really? Because, strangely enough, my experience going through it was that the exhibit was composed of really small, compartmentalized rooms all through the beginning half, until you reached that large dino hall" (fig. 6.6). "And then it was small, compartmentalized rooms all the way out. I kept wondering if this was because dinosaurs are so exciting to most people. Were you thinking that having dinosaurs placed in the center could serve as the edutainment lure that would force people through the other compartments as well? I assume that dinosaurs really turn people on, is that a correct assumption?"

"Oh, yeah, dinosaurs are what turn people on, and dinosaurs are what people tend to remember after the exhibit's over, and want to talk about when you interview them. So yes, dinosaurs; we knew that dinosaurs would be the big draw. Of course, these dinos are all big skeletons, after all, and the sheer size of specimens has to constrain and direct our display decisions. But really, I think the idea behind the design was that the heavy content, on the stuff that people were not familiar with, was the stuff that came first and came afterward, and that's where we really got into the details of the evolutionary process. We talked about material they didn't know a lot about, like the first life forms, and the things that lived in the sea, and that kind of thing. We knew that we were going to wear them out with eight

Figure 6.6. Dinosaur display (wired skeleton) from the Field Museum's "Life over Time" permanent exhibit. Display developers referred to the "dinosaur room," in the middle of the evolution exhibit, as the "candy" for patrons, the lure and the treat that helps visitors digest the more difficult information.

thousand square feet of this kind of stuff, so then we wanted to give them a break. And that's the way we actually talked about it: 'This is where we're going to open things up and give them a break.' 'This is the candy at the end of the meal.' And then we knew that people would be leaving the dinosaur hall with a little bit of energy left, but probably not much, so that's why we kind of did a fairly open space design with a lot of choices in it for the last exhibit hall. People could skip parts of that last hall, and heavy parts of it, if they wanted to. And, sure enough, a lot of people *do* skip the parts that are off the main sequence of rooms. Only half the people, for example, go to the little movie theater at the beginning of the last hall, and another half go into the section on the origin of species and the meaning of species. And only about half the visitors ever go into the section on human evolution."

"Is that right?" I blurted out. "I would have thought that human evolution would have been more interesting to people than anything else."

This area of the exhibit has a humorous display about the origin of *Homo sapiens*. There is a circular hole cut into a wall, and you, the patron, are invited to stick your head inside. When you do this, you find yourself looking into a huge mirror that is reflecting a mural painting of a hominid birthday party; and you (with your head now fused to an apelike body) are the guest of honor, about to blow out the candles (fig. 6.7). The exhibit is fun in that sort of corny carnival way. I was surprised to hear that half the museum-goers never even went into that section.

"Well, people are kind of worn out at that point," Eric explained, "and that section doesn't have big, glamorous skeletons. And some people just plain miss it, I think, as well."

Eric explained that the overall pedagogical blueprint of the exhibit design runs like this. On the one hand, the curators want to tell viewers what happened in the history of life. This is the time-line model of exhibition—start at the beginning of evolution and just move forward through time sequences. But in addition to this, they want to tell us what scientists understand about *why* it happened, the evolution part of it. And that's where the other kinds of spaces came from: the adaptation lab, where the roulette wheels stand, and the mutation lab. In those display spaces the curators tell us *how* scientists know what happened. So in the first part of the exhibit, spaces that tell you what happened alternate with spaces that tell you how scientists know what happened. These were actually called "hows" and "whats" in the design and development stages of the exhibit. The idea was that "how" and "what" would alternate through the end of

Figure 6.7. A humorous interactive display entitled "Happy Birthday *Homo sapiens*" from the Field Museum's "Life over Time" permanent exhibit. Visitors who place their heads through the hole find themselves the guests of honor at a human speciation celebration.

the Paleozoic, until people are sick of them; then the visitor is put into the dinosaur hall, which Eric called the "candy." The dinosaur hall has some evolution buried in it in various places, and one can see family trees of dinosaurs and such, but it is more fun than anything else. Then, in the last part of the exhibit, the layout returns to the "how" and "what" format again, but curators tried to keep things a little bit lighter and more fun.

When describing the behind-the-scenes decision-making process, Eric referred to "we" and "the team" quite frequently. I asked him to clarify this. His actual title at the museum is senior exhibit developer, but I was unclear on the rest of the division of labor. Who makes up the team and how does it work?

In the past at the Field Museum, before the tenure of the current president, a three-pronged team prepared the exhibits. A team that consisted of exhibit developers, designers, and production people would meet several times a week to decide what the exhibit was going to be about. They would go to the curators (researchers such as John Bates) to get ideas and feedback about what scientific themes should be produced, and they'd go to outside experts for feedback as well. The team would also consult with educators within the museum. But now an exhibit developer works on a team with curators and designers; the more traditional production people, the all-rounders, tend to be shut out a little bit more. In fact, for the team that Eric was currently on when I spoke with him, the designers were working in London.

The exhibit that Eric was working on is called "Life Underground," and the Field contracted outside designers for this particular project, which means that exhibit developers were working directly with curators in the meetings in Chicago but then had to send off their ideas to the designers in London. According to Eric, this new process was very awkward. They started out in Chicago by raising some general and specific ideas about what the messages should be for a given exhibit, noting what was available in terms of objects, and developing ideas for interactives and things that might communicate those messages. They put that all together in notebooks and sent it over to London, and the designers over there went through it and then sent their ideas and sketches back, and the Chicago team would react to them. There were regular "fax times" every week.

Eric started rolling his eyes pretty regularly through this last bit, and finished it by saying, "I don't like working at those kinds of distances. That's one reason I'm leaving the Field Museum."

"Oh, really?" I offered casually, all the while drooling for some muck-raking gossip about the new regime. But Eric is too cool for ad hominem attacks. He explained that he is disappointed because the strengths of the old process are being compromised by people who are too distant, both inside and outside the institution.

"The idea of bringing in more outside contractors seems to be very popular now within the upper reaches of the institution," he explained, "and I don't like to work that way. I like the give-and-take of having a team where the designers and the developers are in the same room. And also, I have to say that I'm not really impressed with having the scientists in on weekly meetings, because an inevitable part of any exhibit is that you have to trim the content down to what people will actually look at in the time that they have available to go through the entire museum. But this trimming is a very painful process for the scientists, and they tend to obstruct progress; and I say this even though I myself have a Ph.D. in geology. There are limits to what you can accomplish within a museum exhibit if you're thinking about what you can accomplish with a normal visitor. The scientists want to have way too much information included; it's overkill."

Eric started laughing and impersonating the curators. "'You can't leave *that* out, you can't leave *this* out,' or 'This is *too* important an idea, you'll be confusing people by leaving it out'—that kind of thing is endless. The scientists who consulted with us on 'Life over Time,' for example, browbeat us with their buzzword, *content*. 'You're leaving out the content,' they would shriek. And, of course, the content tends to be what people fill textbooks with, and textbooks these days are a thousand pages long. We like to pretend that students read these information-overload textbooks and get something out of it, but, you know, it doesn't really happen that way. Even in college! But even though many of these curator-scientists have taught college, they haven't caught on to this fact.

"The upshot of all this is that I end up playing an editorial role, always starting with way too much content and editing down and editing down and editing down and seeing what's left. Someday I'd like to do an exhibit where you start with some basic ideas and basic rules, and then you build that to the appropriate level starting from nothing, more or less. Rather than starting with everything and editing down, I'd like to build up from basically where visitors are starting, which is with some very basic ideas that they might remember from school and newspapers and TV."

"Well, what are the presuppositions," I asked, "or the assumptions you have to make about the audience when you're doing an exhibit like 'Life over Time'? I mean, how do you figure out where people are at, so that you can address them at their level?"

Eric explained that the process of gauging the audience is called "front-end evaluation," and that involves developers going out and talking to visitors, posing some very basic questions to find out what visitors know about different topics. The developers ask them how they feel about it and whether they'd be interested in an exhibit on the subject. This method of survey evaluation is encouraged by the National Science Foundation for any project that they give grant money to, and many other foundations that give money now believe that this is the place to start as well. One thing that helped this gauging process in the past, when exhibits were developed with the old-style team method, was that the nonscience people on the team also didn't know anything about the content, and so they naturally asked many questions, including "What the hell are you talking about?" And that sort of question was a healthy thing. *Now* when the team is in a meeting, such as the ones Eric was participating in for "Life Underground," everybody knows everything to begin with, and so it's difficult to get really good estimates of what people *don't* know. It's pretty hard, after all, to remember back to when you didn't know something.

The kind of team that developers put together, its diversity, can determine how well the issues of a given exhibit are handled. For example, the Brookfield Zoo, in Illinois, has a radical approach to this planning stage. They put together teams that are pulled from *every* division of the institution, including keepers, visitor services people, publicity people, and so on. They wind up with even better cross-sections of public knowledge, and this helps resolve potential exhibit problems before it's too late to change them. For example, one problem that crops up in working with visual designers is that they don't really know on a gut level how poorly most people understand design. But production people and developers, and even the curators, can help with that problem because they don't understand design, either. And so when all these people who lack different things try to work together, the things they lack are almost as important as the things they bring to the process. That interplay of expertise and ignorance helps the team create a realistic exhibit.

The next stage, after the front-end evaluation, is called the "formative evaluation." This occurs after the team has been designing some displays and has early stages of production completed. Developers take display prototypes out on the floor, show them to visitors, and see what they react to. The team makes changes accordingly, goes ahead with production, and in the end a "summative evaluation" of the whole finished exhibit is performed. The purpose of the final evaluation is to educate the team better about how the audience reacts to the material. Then when developers tackle the next exhibit, the team will start out with that feedback knowledge.

When I asked Eric about the size of the teams, he explained that "Life over Time" had one lead developer and three subordinate full-time developers. There was one lead designer and ten subordinate designers of various kinds working either part time or full time. Usually each designer takes on one particular part of the project and designs that completely, so not everybody works on the same thing. In terms of production, there was one head of production, various supervisors who went to meetings at particular stages of the process, and sometimes as many as twenty or thirty people working on the exhibit at any point in time. In sum, there were up to fifty people laboring on "Life over Time" for stretches of up to six or seven months at a time.

The whole process probably took five years, but the first year or two consisted in the developers getting up to speed on the subject matter and getting the process started. Once it started to be built, the whole thing was built within about two and a half years. In addition, there was an extra year or so for revisions. The team went back in after the official opening and, based on some things they learned from information sources (including written evaluations, volunteer interviews, and letters), put together a plan for making the exhibit work better for visitors, and then tried to carry that out quickly. But because of all the conflicts and designer time and production time, it wound up taking almost two years. All told, the average lifespan for a permanent exhibit such as "Life over Time" is about ten years.

I explained to Eric that I had spent a little time recently in the ornithology collection and was amazed by the vast kingdoms of neglected specimens that never actually make it to the exhibit halls. The whole range of the collection is staggeringly large. When the museum has such a wealth of specimens, how do developers decide what will be chosen for display? How are specific specimens rescued from museum purgatory?

"We developers," Eric explained, "would have people working for us in collections who would tell us what was available. We'd make a first cut and get permission to actually use these specimens in the exhibit if we wanted to, and we'd give the designer a list of more things than we knew would fit in the case, so they could have some play for the design decisions. Naturally, the designers would wind up dropping out some specimens, and this always irritates the curators to no end. But, you know, if you want the display to look good, you've got to give the designers that kind of leeway.

"The thing to remember about 'Life over Time' is that, besides having some dinosaur skeletons and a few other collections we knew we wanted to use, this was an exhibit that was driven more by ideas and concepts than by specimens. And the fact that we know about the ideas and concepts because of the specimens means that there's a pretty close link when all is said and done. But the ideas are what drove this exhibit, not the collection itself."

Now that Eric was talking about the ideas that drove the exhibit, it seemed to be the proper place to ask him about Alan's letter. I started to explain the distinctive missive to him, and he began smiling and nodding. He knew all about Alan's letter. In fact, he and the other developers had enjoyed it so much that they'd made copies to ensure that they'd always be able to keep Alan tacked up on the talk-back board. I asked him whether or not he thought the gambling displays were unfair biases. He thought not, but then pointed out that it's in the nature of a bias that it be somewhat hidden from the possessor's view.

"People definitely come away from the exhibit," he reflected, "with a sense of message. Alan is not alone in his particular interpretation of the exhibit message, but he does not represent a majority position. There were definite patterns that emerged in our exit surveys when we asked the question 'What message does the exhibit have for you?' Some of the responses were things we had hoped for, and others were unanticipated. A significant number of people walk away from the exhibit feeling that we humans may become extinct, too, if we're not careful. They walk away with the message of human fragility. Another grouping of responses centers on the idea that we humans have occupied a very insignificant slice of the enormous history of the earth; the words *insignificant* and *small* are used frequently. The other trend in these perceived messages was that life is always changing and nothing is permanent. But I hasten to add that most of these message impressions were not reported as negatives by the visitors. People generally had very positive attitudes about these messages.

"Alan's letter," Eric continued, "nicely crystallizes a lot of the issues that we've had to think about while developing 'Life over Time.' We had to think carefully and do front-end studies about where the general public stood on the controversies of evolution. We also did quite a bit of summative evaluation to see how it was coming together for people. We asked people a series of questions in the surveys. One was 'Do you believe in the facts of evolution?' Another was 'Do you believe in God?' And a third one was 'Do you believe that humans are descended from animals?' And we actually did phrase the questions by using the term *belief*, because we do see that for most people it really is a matter of belief. Scientists generally don't like this term, *belief*, but the average person is very comfortable with it. Now, it turns out that something like eighty-five percent of our audience said that they believed in the facts of evolution, ninety percent said they believed in God, and sixty-five percent said that they believed that humans are descended from animals. So basically, if you believe in polls and the way they word questions on this kind of thing, then it's obvious we have a very strongly self-selected audience. Because large-scale surveys tell us that the overall general public is much less supportive of evolution.

"I'm starting to think, however, that these surveys may not phrase the questions correctly. The information we get from the talk-back boards, for example, adds another completely different dimension to our understanding of where the public stands on these issues. And some of the understanding that we're getting has to do with the way that people fit together their belief in God and their belief in evolution, and their belief in whether or not humans are descended from animals. The public puts these things together in ways that are completely different from the way a scientist thinks about such things. The majority of people that give us feedback are able to somehow reconcile contradictory stuff into a combination that they feel comfortable with. The interesting thing about the talk-back boards is that they expose a lot of dimensions of visitor understanding and feeling that wouldn't normally be accessed through survey questions which only interpret things as 'creationist versus evolutionist.' It's just not that simple for most people."

"It's good news for evolution," I offered, "that people are finding ways to make it work with their deep religious convictions, because those convictions don't appear to be disappearing anytime soon. I just wonder whether the price for compatibility is a distortion of our understanding of mutation and selection. Do you think people have to distort the pieces in order to get the puzzle to fit together? Or is there some less gerrymandered road to compatibility?"

"I'm not sure exactly, but when I study the visitor feedback I recognize dimensions that we don't usually think about. Some of the dimensions can be defined by the kinds of questions we asked in the survey, like 'Do you believe that God had a role in making the world it is today?' would be one cognitive dimension. And at one end of the continuum, it's 'No, God did not have a role,' at the other end, it might be standard creationism—God controlled every little detail as of six thousand years ago. But that does not necessarily mean there's not also a separate dimension in people's heads for the role of evolution that may be related—maybe a diagonal access line. I'm not sure. It's like you need a God dimension in this, but you also need an evolution dimension for things other than humans, but then you need an evolution dimension for humans that is separate from the nonhuman conceptual space. Next you need a dimension for whether it's even worth worrying about, and you can be on either end of the scale for whether God has a role *and* say it's not worth worrying about, or it is. And this last bit is a separate thing, it's what runs diagonal to everything else. So I think visitors have this multidimensional way of thinking about these issues. And some of the modalities that people are thinking in are more poetic than anything else. For example, after we put up that little display on the Pope's particular synthesis of these secular and spiritual dimensions, one visitor left a talk-back message saying that 'evolution is the glove on the hand of God.' Isn't that a wonderful metaphor? I mean, you can be an atheist and still appreciate that metaphor, still see how other people are thinking about these things."

Agnostics, atheists, and skeptics tend to think of religious people as "primitive" and "superstitious." Religious people tend to think of agnostics and such as "missing something," as "empty" somehow. The ordinary way of thinking about these things is that religious people affirm a belief in God and nontheists deny that same belief. There's an affirmative and negative judgment regarding the same proposition. And given our understanding of truth and the logical law of noncontradiction, we believe that reality must be either God-populated or not God-populated—it can't be both, even if we don't know the truth right now. But Eric Gyllenhaal's idea of a multidimensional mind is provocative here, because it redraws the map a little bit. It refocuses us away from the traditional attempt to understand how the conflicting propositions or ideas (atheism/theism, or chance/design, etc.) can fit together without contradiction. It takes us away from that disembodied, decontextualized approach and returns us to the people themselves. Eric's approach gets me to thinking, not about abstracted beliefs and propositions, but about the person and the context of his or her life, which grounds those beliefs. When I reflect on that, I no longer think that Alan and the agnostic (someone like me) are talking about the same propositions; I no longer think that our multiple dimensions are intersecting in the same places. Certainly we all share terms in our different discourses, but we do not organize our lives according to the same deep pictures.

The NOMA proponents (i.e., those who assert that religion and science are nonoverlapping magisteria) claim that the terms and ideas of the two domains are not comparable; they're apples and oranges. And the reason for this lack of comparability is that one domain has a language and way of thinking that is literal, objective, and factual, while the other domain is metaphorical, subjective, and value-based. The fact/value split creates the nonoverlapping magisteria.

I, on the other hand, submit that while there are many terms and ideas that cannot be compared between domains, there are also many significant overlaps. Some of the overlaps are fruitful, and some of them are not. I think Eric Gyllenhaal, with his model of a multidimensional mind, would agree with me here. Overlaps certainly make things messier, but the situation is not hopeless. Facts and values, or the mundane and the spiritual, are not distinct domains. They permeate each other, and not in some mystical everything-is-one sort of way. I mean that the practice of doing science is saturated by human values, and that the religious viewpoint is very much an interaction with the factual world. We can intellectually separate these things, but in reality they are intertwined like form and matter. And it is ironic that Gould formally states a strong fact/value distinction when his own excellent writings on "scientific" racism and bogus intelligence testing are saturated with and motivated by truly noble values. His own opposition to social Darwinism and sociobiology is heavily influenced by his liberal value system. Yet the NOMA principle seems to rule out this interpenetration of facts and values.

While Gould likes the metaphor of two separate territories, and Gyllenhaal likes the metaphor of intersecting dimensions, I prefer the Wittgensteinian metaphor of pictures. A picture is a composition of different forms and patterns. Sometimes the forms are related to each other in a concordant way, sometimes not. Sometimes small parts of the overall composition seem to coordinate well, while other parts hang together in a haphazard way. The individual human mind and life are like this. The gestalt composition of our ideas, emotions, beliefs, background assumptions, and experiences cannot be completely sifted into two realms, facts and values. At bottom, Alan and I have two very different pictures that govern our lives. In my picture, I can't get the random-mutation pattern to coordinate with the beneficent-designer pattern. The combination clashes too much for me, but Alan can pull it off. This difference has to do, in large part, with what other forms or patterns (i.e., ideas, experiences, or emotions) these two are attached to or mixed with. Isolating one feature of Alan's picture, or my picture, and showing all the ways it doesn't make sense can be a legitimate and fair exercise in some cases, but it also fails to recognize how that feature is connected to other important ways of living and being in the world. For Alan, the belief in a nonrandom nature is connected in an important way to a feeling of gratefulness or a general appreciation of life itself, whereas my feelings of appreciation are connected to completely different parts of my picture. A person's organizing picture structures the way that person thinks, imagines, and behaves. And while that picture might include a very large section that is given over to a pattern called logic, we should not be so naive as to think that the picture itself is constructed on the foundation of logic.

When the believer in randomness and the believer in design go head to head, they bring with them their respective pictures. If ontological beliefs actually existed in an isolated way within people's minds, then straightforward refutations would actually work; people would actually hold up their belief to specific tests, see that a specific belief doesn't accord, and happily discard it. But because the belief is tied to so many other things, some of which may be very precious commitments, people will hold fast to the belief and reinterpret the tests instead. This happens as much in science as it does in religion.

One of the positive outcomes of this metaphor of pictures is that people who disagree with you cannot be easily dismissed as either stupid or deceptive. Some deep disagreements between people may have more to do with the picture contexts than with the isolated terms under discussion. Does this mean that there is no shared vantage point from which to critique each other's beliefs or even pictures? Are Alan and I simply two ships passing in the night, two people holding incommensurable pictures? No.

Two things are clear about the pictures that organize our respective lives: They are revisable, and they have many features in common with other people's pictures. While a person's core picture of life is built up by gradual accretion and is slow to change, it is not immovable. New experiences, persuasion, and logic are always tinkering with the picture. Second, our common humanity gives us very similar raw materials for our pictures (needs, wants, hopes, etc.), and this allows each of us to gain some entry into another person's mode of life. My picture and Alan's picture are deeply rooted things, but they are not unchangeable and they are not totally incommensurable.

So, with regard to the question of who's right, I think it comes down to the issue of what's at stake. If Alan and I are having a lively debate over coffee somewhere, and he's advocating for design and I'm championing randomness, then there is probably not much at stake; if we both walk away from our exchange with unchanged pictures, then it's no big deal. If belief in a designer God helps the Alans of the world to get through the day, or if it inspires them to acts of charity, then it seems obnoxious (and wrongheaded) to dismiss this ontological commitment on the grounds that empirical data fail to prove the truth of the assumptions behind it. But if the same belief devolves from healthy inspiration to dangerous hostility toward alternative ontologies—say, if Alan is lobbying the local school board to censor the parts of biology textbooks that discuss random mutation—then much more is at stake in this second scenario, and I'm going to work very hard to get Alan to revise his picture.

It is, in fact, possible to be trapped by one's own picture. A person can become so attached to an ideological picture that he ceases to be open to the lessons of his own experience. But I would not wish to argue that being trapped by one's own picture is harmless unless something important, such as personal or political freedoms, is at stake, for to argue in this way is essentially to say that the Alans of this world are stupid but can go on being stupid as long as they don't hurt anyone. Perhaps the proper way to understand this scenario is to say that the rightness or wrongness of these sorts of beliefs can be assessed only in a particular context; that is, Alan did not actually become trapped by his own picture (did not actually become wrong) until he used his belief in design as part of a specious moral argument for censorship.

In this particular case, I have argued that the correct metaphysical posture is the chance-based (nonteleological) ontology that the Field Museum purportedly conveys in its exhibit. As a matter of sheer principle, this nonteleological position best reflects the empirical data of evolution; and Occam's razor, together with the problem of animal suffering, discounts the need for or the reasonableness of positing a personal designer at evolution's outset. But I then switched gears and tried to point out that these contentious beliefs, propositions, and commitments are more than disembodied principles. They must be understood in the context of the whole person. This move pragmatizes the competing truth claims, tying them to larger psychological, aesthetic, and social dimensions. What ideas and practices stand or fall with the particular ontological commitment? This pragmatic move (admittedly, diplomatic in nature) allows one to have a charitable and tolerant attitude toward others' commitments when those commitments remain personal. But one can also justify a fight if such commitments start to become public mandates. Alan's commitment to a designer, for example, is not wrong when it is giving him personal inspiration, but it is wrong if it seeks to legislate the lives and thoughts of others—if it seeks to narrow rather than broaden intellectual options.

Unfortunately, while this diplomatic move might make agnostics and atheists more comfortable, it sets up new problems for the Alans of the world. Consider that Alan is inspired, heartened, and hopeful *because* he believes that God designed the cosmos. But can a belief that gives personal inspiration *because* it purports to be about the physical world continue to give personal inspiration if its claim about objective reality becomes forfeited? It seems unlikely, and yet something like a forfeiture seems to be required when I argue that proponents of a cosmic designer should recognize the personal and subjective character of their spiritual convictions. How to resolve this is unclear. We are forced again to ask, what is at stake in the conflict?

This leads us back to our beginning. What is at stake at the Field Museum roulette wheel? Who's betting, and what are they wagering? I argue that the stakes in this ideological clash do not have much to do with morality; Alan, on the other hand, claims that they do, and furthermore that the notion of a chance-based universe is morally bankrupt. If the NOMA proponents are correct, then there's no real way to resolve this disagreement because it falls into a domain that does not deal with the empirical universe. The implication of NOMA is that any revision of thinking in this metaphysical magisterium must occur by conversion through mystical intuition, because revising a position is normally based on exposure to new information (new data), but by definition this new information cannot be empirically given.

To be precise, however, moral issues are issues regarding human action. We can theorize and philosophize all we want about morality, but moral character is a result and a cause of actions; intentions and ideologies may inform actions, but it is the action that counts. Reading books, studying scriptures, having philosophical conversations about truth-telling behavior—all these may inform one's moral life, but one's moral status is actually constituted by one's lying or truth-telling behavior. The same can be said of the other virtues and vices. In morality, thoughts are not as important as deeds.

Given this fact, it would be rather unintelligible to argue that a person could hold an immoral ideology but behave morally. In other words, the terms *moral* and *immoral* (or any other parallel terms) apply only to a person's actions. Ideologies, properly speaking, are neither moral nor immoral. So the only way to know whether believing in a random universe is morally bankrupt is to study and assess the actions of people who hold that belief. To engage in such a study would be to engage in an empirical study. Admittedly, this would be a difficult study to make, but it is not impossible, nor is it incoherent. Alan's hypothesis, that belief in roulette ontology is immoral, is actually testable, as is my counterhypothesis. We may not have done enough empirical research to know which of us is correct, but it is clear that some resolution is eventually attainable. And the domain of that resolution will be experiential.

CHAPTER 7

DRAMA IN THE DIORAMA: THE CONFEDERATION OF ART AND SCIENCE

THE MUSEUMS THAT WE'VE STUDIED throughout this journey reveal the tremendous diversity of goals and motives for collecting and displaying elements of the natural world. Yet underneath all these various constructions of nature, there has been a continuous dialogue between image-making activities and knowledge-producing activities. Unlike texts, natural history museums are inherently aesthetic representations of science in particular and conceptual ideas in general. The fact that a roulette wheel at the Field could touch the central nerves of our deep metaphysical convictions is an indication of a museum's epistemic potential. After spending long stretches in many natural history museums, one begins to see that a display's potential for education and transformation is largely a function of its artistic, nondiscursive character. Three-dimensional representations of nature (dioramas), two-dimensional and three-dimensional representations of concepts (such as the roulette wheel), and visual images generally are not just candy coatings on the real educational process of textual information transmission. This chapter explores how and why visual communication works on museum visitors. And this requires an examination of the more general issue of how images themselves can be pedagogical, an issue that extends from da Vinci's anatomy drawings to the latest video edutainment technology. These issues lead to a survey of some of the most recent trends in museology, followed by some reflections on the museum at the millennium.

The artistic features of an exhibit directly address the all-important but nebulous emotional aspects of the learning process. Educational psychologists have generally distinguished three different domains of learning. Psychologist Benjamin Bloom formulated a well-known tripartite taxonomy of educational objectives (usually referred to as Bloom's taxonomy) that distinguished between affective, cognitive, and psychomotor areas of learning. Affective learning pertains to feelings, emotions, attitudes, and values. Cognitive learning pertains to remembering, combining, and synthesizing information. And psychomotor learning refers to muscular skills, manipulation, and coordination. Obviously, these domains interact with each other and do not exist independently, but museologist Lisa Roberts contends that museums are uniquely and profoundly affective. She points out that "it is the nature of our institutions—multisensory, three-dimensional, interactive—that they should appeal so strongly to that part of the brain concerned with space, image, affect." Images and dioramas, for example, are extremely good methods for establishing mood, and experimental psychologists have documented significant connections between mood and memory. How you feel during a specific experience becomes a crucial aspect of your memory of that experience. If that mood is evoked at a later date, it can trigger the memory of certain details that were absorbed during the original experience.

We live in an age when the stereotype of the artist is primarily romantic. We've placed the brooding, self-absorbed expressionist on the pedestal of fine art and forgotten that other forms of sensual communication are equally artistic. Everywhere you cast your eye in a natural history museum, for example, it lands on an artwork, a carefully crafted, aesthetically thoughtful creation that is both born of and evokes inspiration.

Our era has been duped by C.P. Snow's infamous dichotomy of "two cultures." The dichotomy itself is ancient and comes in and out of vogue, but the most recent incarnation stems from Snow's 1959 book *The Two Cultures.* Snow suggested that the sciences and the humanities were really two different cultures. This view maintains an artificial divide, with scientists thinking of art as superfluous (or, as Francis Bacon referred to it, "feigned knowledge") and artists thinking of science as glorified stamp collecting. But the natural history museum clearly puts the lie to this two-culture simplification.

In the gigantic central hall of the Field Museum is a very famous display of two huge African elephants (fig. 7.1). The life-sized taxidermy display is the creation of Carl Ethan Akeley (1864–1926), a remarkable artist and naturalist who, together with William Hornaday (see chapter 1), revolutionized museology (fig. 7.2). By the early 1880s Akeley had moved away from the lifeless stuffing techniques of his contemporaries (the upholstery techniques), and in 1885 he was given the opportunity to mount P.T. Barnum's famous circus elephant, Jumbo. By constructing a lightweight wooden armature, Akeley brought Jumbo back to life. And he brought these revolutionary action-pose techniques with him when he joined the Field Museum (shortly after its origin) in 1895. Akeley's artistic powers were heightened by his firsthand studies of animal anatomy and animal behavior during many African safaris. In addition to the breathtaking elephant mount, almost all of the Field Museum's African mammals still on display were done by Akeley's artistic hand. Akeley, and most other natural history museologists, stand firmly with one foot in each of C.P. Snow's two cultures.

So sensitive was Akeley's ability to communicate through three-dimensional media that his first small-scale bronze sculpture, entitled "The Wounded Comrade," won him admittance into the National Sculpture Society in 1913. The artwork, an eleven-inch-high elephant sculpture, was simply an extension of Akeley's museum work; art and science were always wedded for him. Akeley summarizes how his own work is simultaneously scientific and artistic communication:

> I have many tales to tell of African animals, but only so much can be told in a single taxidermic group; and the making of such a group—the collecting of specimens and data, the modeling of the animals, the construction of the manikins, the painting of the background, the production of the accessories—all this preparation will take months and years. . . . However, I am able to put into small bronzes many records of the life habits of animals that cannot be put into taxidermy. . . . My *Wounded Comrade* portrays the fraternal spirit of elephants which others, as well as I, have noted when two or more elephants in a herd rush in to lift and, if possible, to carry to safety a wounded companion.

Akeley's scientific artwork led me to think about the relationship between the aesthetic and didactic dimensions of the museum. Remember that Robert Hooke warned fellow scientists of the Royal Society to be very careful when using pictures to educate people, because images can "divert and disturb the mind, and sway it with a kind of partiality or respect" (chapter 1). In Hooke's mind, images can be so powerful that their subjective appeal should not be mixed with the realms of objective truth. The alienation of reason and emotion during the scientific revolution led to a bifurcation of theory and perception. But ironically, the original Greek word *theoria* meant "seeing" or "viewing" as well as "contemplation." Imagery, for many Enlightenment scientists, was the medium of mere spectacle and needed to be subordinate to written text and mathematics.

Figure 7.1. Two African bull elephants created by master taxidermist Carl Akeley. The elephants occupy the great room of the Field Museum. (COURTESY OF THE FIELD MUSEUM, NEG. NO. GN79303)

Figure 7.2. Carl Ethan Akeley (1864–1926) in the field during one of his many African safaris. (COURTESY OF THE FIELD MUSEUM, NEG. NO. CSZ6167)

Throughout Western culture, from Pythagoras to the present, there has been a traditional hierarchical view of cognition, with sensual perception at the bottom, linguistic thought above that, and mathematics at the crowning position. Leaving aside the whole tempting business of which is higher and which lower, let's try to appreciate how the visual arts and the sciences can, and have, become allies. In examining some specific cases, we may come to see that there is a great deal of visual thinking in the scientific and lay communities.

In Neil Campbell's 1999 textbook, entitled simply *Biology*, he states, "Biology is a visual science, and many of our students are visual learners." For this reason, he claims, a textbook must contain extensive imagery. Biology and its scientific progenitors have always been very descriptive enterprises. Medicine, natural history, and anatomy have always sought to condense information into image formats, but, more important, they have always had uniquely visual subject matters. Whether one is talking about cladograms, microscopy images, or anatomical maps, the learning of life science is largely visual.

At the entrance of Philadelphia's Mutter Museum (perhaps the only collection more macabre than Hunter's), the curators have included a passage from S.H. Daukes's 1929 book *The Medical Museum*. While Daukes refers specifically to the medical field, his general point can be extended to the other life sciences as well. He states, "In clinical work, the eye is the main avenue for diagnostic information . . . in medical education, again, the eye must also take first place; it is from the wealth of material stored as visual impressions that the expert draws his comparisons and conclusions. A graphic method of museum demonstration is calculated to sharpen and develop this sense and provide a preliminary foundation upon which subsequent experience can build."

Graphic presentations, both two-dimensional and three-dimensional, that can be collected and displayed serve the professional as much as the lay person. An artistic presentation can make scientific matters more clear, and, vice versa, causal scientific analysis can render aesthetic expressions more compelling. Artists can benefit from the same specimen that sheds light on scientific issues. Biological specimens are functionally promiscuous, and a great example of this can be seen in William Hunter's "muscle men" statues.

John Hunter's older brother William (1718–1783) was appointed chair of anatomy for London's Royal Academy of Arts in 1768. William, who first tutored his younger brother John, had been a well-known anatomist in London, and the Royal Academy was privileged to add him to their elite company. According to the academy's *Instrument of Foundation*, the chair of anatomy was to "read annually six public lectures in the schools, adapted to the Arts of Design; his salary shall be thirty pounds a year; and he shall continue in office during the King's pleasure." In these public lectures William explained to his students that "a very correct knowledge of Anatomy must be of the utmost consequence to every artist who is to make resemblances of the human Body. It enables him to observe and distinguish clearly all the variations of form, because it explains their causes."

During some of these lectures William would demonstrate muscle anatomy on a life-sized plaster statue that he had brought to the academy. The story of this muscle man is strange but somehow consistent with other Hunterian events. It turns out that in the early 1750s his brother John was awarded eight bodies for dissecting; these were convicts who were executed at Tyburn. William was, at this same time, giving anatomy lectures to the Incorporated Society of Artists and needed a fresh corpse. William's group chose a cadaver from John's cache and shuttled it back to their apartments, but when William found the body so fresh—not yet stiffened—he hit upon a novel idea. Hunter placed the body into a display position (arm raised, leg posed) and simply waited for rigor mortis to set in and freeze the fellow into place. Then William and his students (and possibly John) worked all night to strip off the flesh and expose the muscles below. They then made a mold of this muscle man and cast the plaster statue (fig. 7.3). Clearly, John was not the only member of the Hunter family collecting odd specimens. In effect, a real man was chiseled out of his skin to make a muscle-man statue so that artists could learn to reskin and reclothe him into an accurate "resemblance" of a real man.

Besides the Hunter brothers' penchant for collecting and employing three-dimensional specimens, they also sought to preserve their most evanescent subjects by employing talented painters and engravers. Drawings, paintings, and etchings of specimens have, until recently, formed a kind of metamuseum or epicollection, a graphic account of nature's diversity and nature's organization that can function as a portable museum. And just as S.H. Daukes claimed that "the eye must take first place" when learning about life processes, a good scientific illustration was worth a thousand words. A truly sensitive illustration possesses that elusive dual function of didactic instruction and affective impact. The artistic representation of a specimen, unlike the indiscriminate camera, must also walk the razor's edge between an uninformative abstract schematic, on one hand, and an idiosyncratic, hyperexact representation, on the other.

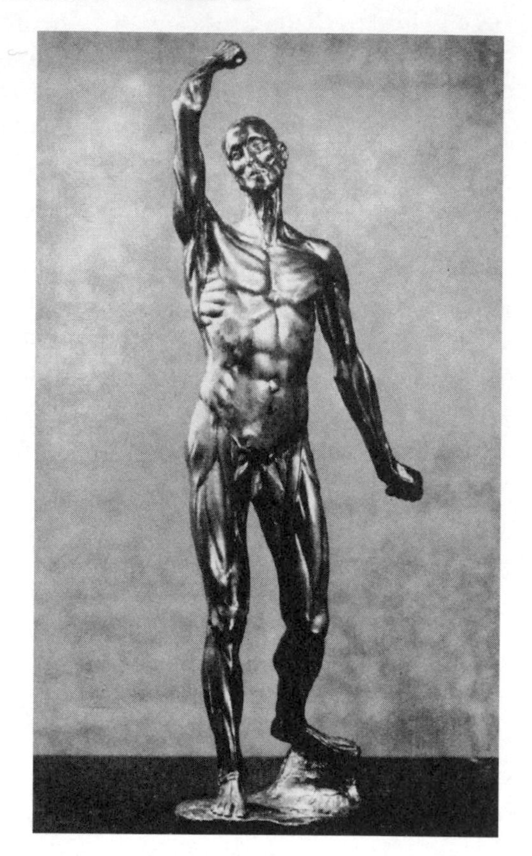

Figure 7.3. A small bronze anatomical *écorché* by Edward Burch, based on William Hunter's cadaver castings. London, Victoria and Albert Museum.

One of the greatest scientific illustrators who ever graced the planet was on the payroll of both Hunter brothers during the eighteenth century. His name was Jan Van Rymsdyk. Rymsdyk belongs in the same lofty category as da Vinci and Vesalius, that weird category wherein the distinction between fine art and science simply does not exist.

The life and death of Jan Van Rymsdyk are shrouded in mystery. The little that we do know is both fascinating and tragic. We're almost certain that he was born in Holland, but we do not know an approximate date of birth. For that matter, we're also not sure when he died, though it must have been sometime in the 1780s. He mysteriously materialized in London in 1750 as one of William Hunter's illustrators. Before these astounding drawings for Hunter, Rymsdyk must have had some significant training or apprenticeship, but we have no records and no surviving works prior to 1750. The drawings that Rymsdyk did for Hunter that first year were anatomical studies of a woman who had died while pregnant. William Hunter was one of the principal players in the developing obstetrics movement, and while he supervised his brother John's dissection of the gravid (full) uterus, Rymsdyk sat nearby, carefully drawing the rendered flesh. This project eventually evolved into William Hunter's magnum opus *The Anatomy of the Human Gravid Uterus* (1774), a project that Rymsdyk almost single-handedly illustrated over a twenty-two-year period (fig. 7.4). Over that span William Hunter procured more than twelve recently deceased women, in various stages of gestation, and employed Rymsdyk to graphically capture the evanescent abdominal dissections for posterity.

At first it is difficult to look at these drawings, or perhaps it's difficult to look away. They are works of genius, but they're not beautiful, nor are they pleasant. The rendering is truly exceptional, in the sense that Rymsdyk conveys a palpable sense of the subjects' weight and mass. The shading gives the forms a sculptural quality, and his facility with the limited black-and-white medium still conveys the subtle differences between opaque and translucent tissues. On top of this, the sheer detail is utterly mind-numbing. Among the many painstaking plates, one finds images of the fetus in situ at term, a retroverted womb in the fifth month of pregnancy, a fully opened womb containing a fetus at the beginning of the fifth month, a frontal view of the womb cavity with child removed, and so on.

It is these images, more even than the text, that secured William Hunter's reputation. Moreover, Rymsdyk's eye was so good and his hand so true that the *Gravid Uterus* drawings are still consulted by medical students to understand the embryological phases of the

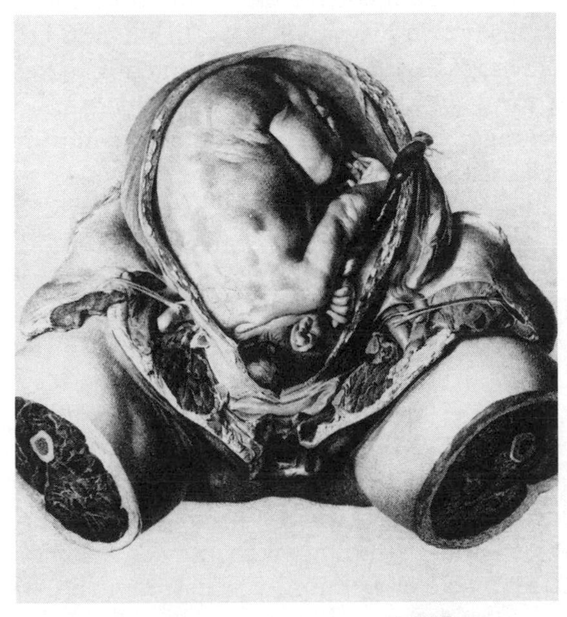

Figure 7.4. One of Jan Van Rymsdyk's drawings of a dissected pregnant cadaver, from William Hunter's *The Anatomy of the Human Gravid Uterus* (1774). The woman's legs have been removed at the thighs, and the child is revealed in situ. (FROM IMAGES OF SCIENCE, BY BRIAN J. FORD, © 1992 BY BRIAN J. FORD. USED BY PERMISSION OF THE BRITISH LIBRARY)

uterus and the shifting of adjacent organs. Yet, amazingly, William Hunter never even mentions Jan Van Rymsdyk in the preface of *Gravid Uterus*. Hunter's omission of credit to Rymsdyk is still not understood, but it was not a lack of appreciation that led to the neglect. William not only saw the value of hiring him over a period of more than twenty years, but even went so far as to purchase the drawings that Rymsdyk did for another anatomist.

The pedagogical uses of Rymsdyk's uterus drawings cannot be overestimated. In the late eighteenth century, understanding of the human female anatomy was very incomplete. The ancients, Aristotle and Hippocrates, believed that the human uterus had two chambers (males developed in the right half, females in the left). The fourteenth-century Italian anatomist Mundinus claimed that the uterus contained seven cavities. And while da Vinci was much more accurate regarding this parturient subject matter, his notebook drawings of female anatomy were not available to practitioners until the nineteenth century. From the Renaissance to Hunter's day, slow and piecemeal progress was made by anatomists such as Gabriel Fallopius (1523–1562) and Abraham Vater (1684–1751), but nothing compared to Hunter and Rymsdyk's thorough and illuminating compendium of gestation.

Medical students, midwives, women, and children all benefited from Rymsdyk's artistic creations. When visual images are actually saving people's lives, it's hard to maintain that they're merely ornamental. And Rymsdyk's unparalleled skill also reached across the Atlantic, helping to place American medicine on a more solid foundation. This is because Rymsdyk, in his early days, sold a handful of his impressive pastel drawings to an anatomist named Charles Nicholas Jenty. Jenty, in turn, sold the dissection drawings to John Fothergill, who passed along the pastels to Philadelphia educator William Shippen Jr. (1736–1808). Shippen used the drawings to teach medical students anatomy, and the pastels are still located today in Pennsylvania Hospital. Students came from far and wide to study these excellent drawings, which were referred to as the "Van Rymsdyk crayons." And, as historian Florence M. Greim puts it, these drawings "played a basic part in early medical teaching in the United States."

The somewhat tragic aspect of Jan Van Rymsdyk is that, for all his scientific acumen, he wished to become a portrait painter above all else; and for lack of dependable patronage, he was forced to live hand to mouth (supporting his son, Andrew) in the company of decomposing and fetid cadavers. A very small sampling of Rymsdyk's work for William Hunter will suffice to give us a glimpse of a day in the life of this artist (keeping in mind that he was simultaneously working hard for other anatomists during this same period). The Hunterian Museum in Glasgow, Scotland, which still holds many of Rymsdyk's works, contains some fifteen drawings of abnormal fetuses and monsters, fifty-seven drawings of diseased organs, seventeen of diseased and normal bones, six of the male reproductive system, a great many images of the gravid uterus, and many miscellaneous dissections.

It was Rymsdyk's passion for natural history that led him to become friends with William's brother John. Like John Hunter, Rymsdyk's curiosity extended beyond medicine to the wider aspects of natural history; though as a visual artist, he focused more on the structures than the functions of life. Rymsdyk grew to resent William's lack of formal acknowledgment and preferred the company and work of the younger brother, whose book *The Natural History of the Human Teeth* (1771) Rymsdyk illustrated. While John Hunter was transplanting teeth into rooster combs, his friend Rymsdyk was nearby, visually rendering dentition specimens in his wondrously detailed drawings (figs. 7.5, 7.6). Rymsdyk's Zen-like immersion into his subject matter, whatever the subject matter happened to be, prepared him for his most ambitious and personal work.

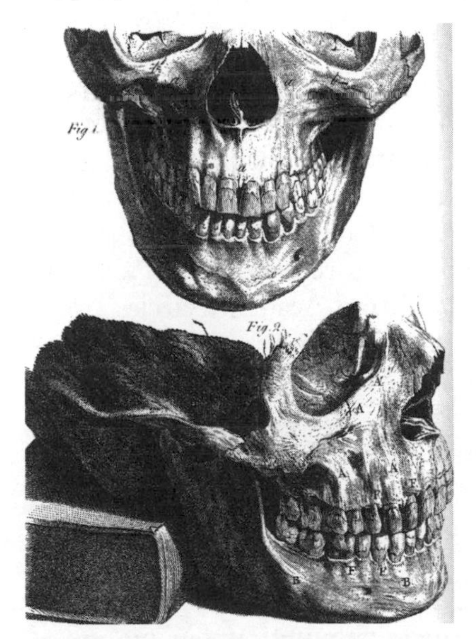

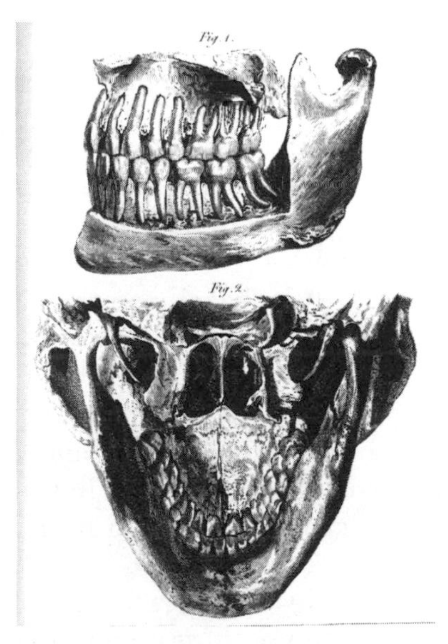

Figure 7.5. A front view of the upper and lower jaws of an adult with a full set of teeth. One of Van Rymsdyk's images for John Hunter's *The Natural History of the Human Teeth*, London, 1771.

Figure 7.6. A side view of the upper and lower jaw, with teeth exposed in their sockets, and a view of the closed mouth from underneath the skull. One of Van Rymsdyk's images for John Hunter's *The Natural History of the Human Teeth*, London, 1771.

Remember that the British Museum (offspring of Sir Hans Sloane's collection) opened its doors in 1759. Starting in 1762, Jan Van Rymsdyk and his son, Andrew (an outstanding apprentice to his father's trade), began to spend long hours at the museum. Rymsdyk spent over fifteen years at the museum, engaged in a secret therapy that he conducted in between anatomical commissions. He unleashed his peerless observational skills on the museum specimens that he encountered, and in 1778 he published a book with the hefty title *Museum Britannicum, being an exhibition of a great variety of antiquities and natural curiosities, belonging to that noble and magnificent cabinet, the British Museum. Illustrated with curious prints, engraved after the original design, from nature, other objects, and with distinct explanations of each figure, by John and Andrew Van Rymsdyk, pictors.* And after this bulging title, Rymsdyk includes a lyric from *Aesop's Life* that reveals his own attitude toward naturalistic drawing and manifests his hostility toward his overly academic contemporaries:

> No more you learned Fops, your Knowledge boast,
> Pretending all to know, by reading most,
> True Wit, by Inspiration, we obtain,
> Nature, not Art, Apollo's Wreath must gain.

The Rymsdyks' "museum on paper" contains many beautifully drawn images, each of which embodies the patient care so obvious in the Hunterian drawings. Rymsdyk must have felt liberated by his *Museum Britannicum* work, because in the text (which meanders strangely) he absolutely unloads on William Hunter. Class differences may account for some of William's refusal to promote the laborer Rymsdyk among the haughty Royal Academy circles (John Hunter, having less of a "womanly" classical education, as John himself called it, always saw himself as a salt-of-the-earth type, and this may have contributed to his and Rymsdyk's solidarity). In any case, it was a long time coming, and Rymsdyk finally lambasted William (referring to him as "Dr. Ibis") for his neglect and lack of support. After likening William to a cur, Rymsdyk states:

> It is a great comfort to me that he is *Alive*, and will see the above [insult], for . . . it is a dishonest mean cunning, in making one self a great Man with other People's Merit. (This is what the Country People call *Reaping without Sowing*.) Pray now, as you was very lucky, and did well in the World, what prejudice did I ever do you, why should you discourage me as a Painter; was I not to live too?

There's something heart-wrenching about this open display of bitterness, and something even more depressing about the fact that the *Museum Britannicum* didn't really make Rymsdyk any money, didn't really liberate him at all. To top it all off, both father and son were dead only a few short years after publishing the stillborn book. But despite the personal tragedy, the *Museum Britannicum* is a true treasure, a tour de force.

What makes the work so fascinating, besides its affective strength, is its position in the development of museology. The paper museum contains some thirty separate image plates, and taken together they comprise a weird antipodal cabinet to John Hunter's museum. Consciously or unconsciously, Rymsdyk's *Museum Britannicum* is the polar opposite to John Hunter's system-based collection, the other part of the split personality. Rymsdyk not only ignores the organizational themes that are emerging in life science (emerging in large part because of Hunter), but actually ignores the modern division of natural versus artificial objects. Rymsdyk engages in a bizarre and defiant return to the disorganization of the earlier curiosity cabinets, even going so far as to include (without irony) fanciful or inauthentic objects. He does not even pretend to appeal to a specific audience of naturalists; he does not try to convey a theoretical slant in the choice of his subjects or in their placement. It is a paradoxical book in the sense that it is at once totally universal (appealing to all audiences equally) and totally particular (each object is an idiosyncratic individual, rather than a specimen). A few examples will make things obvious (figs. 7.7, 7.8, 7.9).

The exhibits are composed as "Tables," and inside each table, one or more "Figures" might reside. Table IV contains Figure 4, "a brass spear-head from Scotland," and Figures 5–9 are arrowheads, four being of brass and one of flint; Table V contains fifteen different ova, or eggs; Table VI contains, among other things, a Spanish dagger; Table VII includes different spider nests; Table XIII shows a brick from the Tower of Babel; Table XVII illustrates all manners of gaming dice; Table XVIII, amulets or charms; Table XXI shows "Lachrymatories" or tear vials, to contain the tears of weeping friends, and which are buried with the dead; Table XXX contains four figures of birds, including hummingbirds and birds of paradise. Rymsdyk's mysterious project is so catholic that in the same drawing he includes owl's eggs and a hairball that was found in an ox's stomach (figs. 7.10, 7.11).

The *Museum Britannicum* is an empiricist celebration of the idiosyncratic truth of visual reality. And while it diverges strongly from his friend John Hunter's project of organization along theoretical

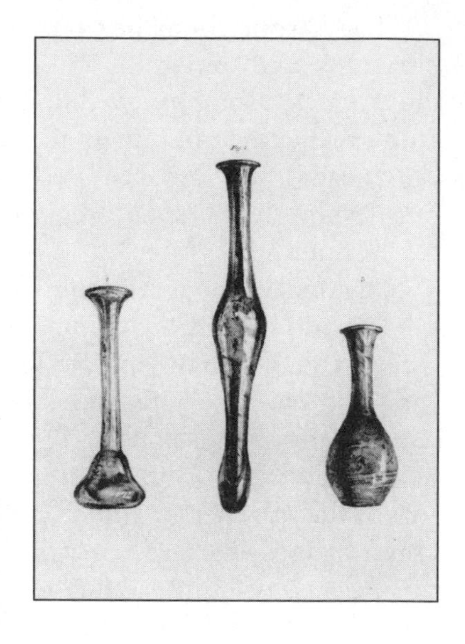

Figure 7.7. Table XXI, "Lachrymatories," or tear vials, from Jan and Andrew Van Rymsdyk's *Museum Britannicum*, London, 1778. (VISUAL MEDIA SERVICES, THE GETTY RESEARCH INSTITUTE)

Figure 7.8. Table XI, "Spider's nest and miscellaneous weaved fabric," from Jan and Andrew Van Rymsdyk's *Museum Britannicum*, London, 1778. (VISUAL MEDIA SERVICES, THE GETTY RESEARCH INSTITUTE)

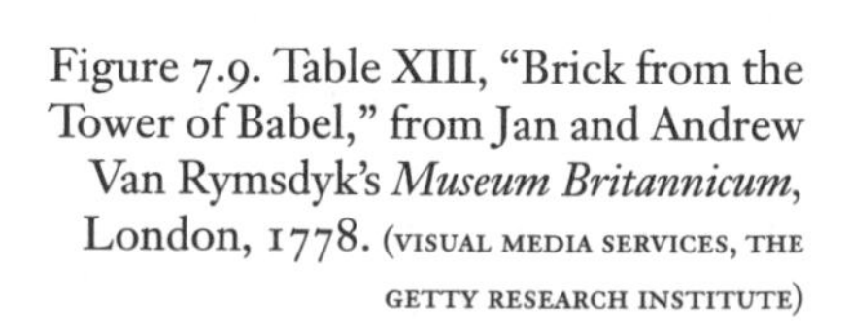

Figure 7.9. Table XIII, "Brick from the Tower of Babel," from Jan and Andrew Van Rymsdyk's *Museum Britannicum*, London, 1778. (VISUAL MEDIA SERVICES, THE GETTY RESEARCH INSTITUTE)

Figure 7.10. Table XXX, "Aves, Birds; Bird of Paradise," from Jan and Andrew Van Rymsdyk's *Museum Britannicum*, London, 1778. (VISUAL MEDIA SERVICES, THE GETTY RESEARCH INSTITUTE)

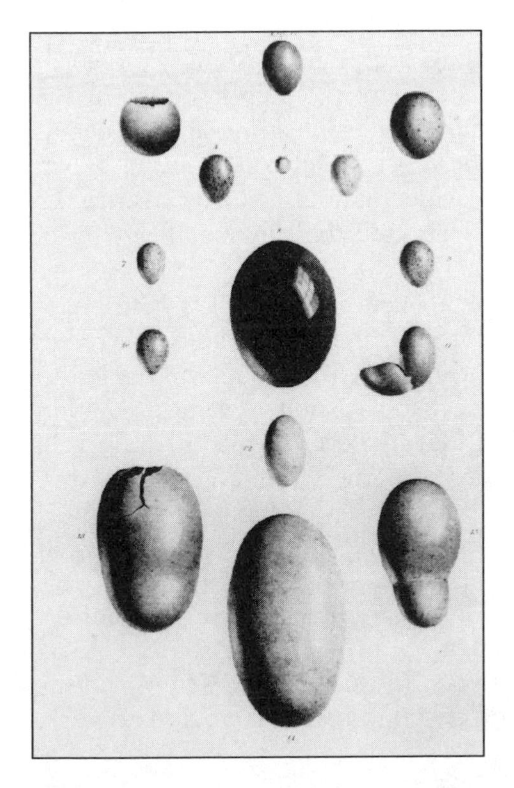

Figure 7.11. Table V, "Various Ova, Eggs," from Jan and Andrew Van Rymsdyk's *Museum Britannicum*, London, 1778. (VISUAL MEDIA SERVICES, THE GETTY RESEARCH INSTITUTE)

lines (functional physiology, in this case), it shares a deeper politico-philosophical sympathy. Rymsdyk makes it clear in his *Museum* text that the elite academic artists (presumably the Royal Academy pals of William Hunter) looked down their noses at the Rymsdyks, claiming that their work was too minutely detailed. Like John Hunter's rejection by his nonexperimental contemporaries, Rymsdyk had no patience for the merely bookish approach to knowledge. To Rymsdyk, these "learned fops" at the Royal Academy spent too much time reading about nature and not enough time getting firsthand experience of it. Rymsdyk's image of communing with nature is not that of the Romantic hero, standing with windblown hair and noble staff on a majestic mountaintop or rocky beach. It's a guy hunched over a little object, straining to get his eye and his mind into every tiny crevice; it's the sublimeness of the small. Jan Van Rymsdyk would have agreed with Mies van der Rohe's famous claim that "God is in the details."

Rymsdyk was a quintessential visual communicator. A late-twentieth-century scientific illustrator, Phyllis Wood, explains that "the artist must be aware of the viewers' level of knowledge and must relate the message in a logical sequence without confusing them with too much or too little information. The artist must be knowledgeable about the subject and draw in a selective and interpretive manner, not photographically recording everything." Rymsdyk, accordingly, knew how to negotiate his audience. When he drew for the Hunters and their medical audiences, he distilled fine-grained details and archetypical patterns into the most useful combination. When he drew for himself and for the wider natural history audience, he torqued up the specificity and celebrated the sublime affectivity of nature.

Many of the important nineteenth-century naturalists were amazing artists as well. Ernst Haeckel (1834–1919), sometimes referred to as the German Darwin, was a brilliant artist, though he tended to represent a very stylized and idealistic version of nature. Darwin himself always lamented the fact that he never learned to draw well, admitting that such a shortcoming hindered his natural history work when out in the field. Richard Owen, the anatomist who curated Hunter's collection and developed the Kensington museum, was an excellent artist who used visual images to communicate ideas (like his vertebrate archetype) and to aid memory. He was the chief architect of the three-dimensional re-creations of dinosaurs at the Crystal Palace outside London in the 1850s. In this sense, Owen provides a nice example of one of the other functions of museological artwork, namely, persuasion.

Owen, working closely with an artist named Benjamin Waterhouse Hawkins, had argued for a very specific three-dimensional representation of the mysterious iguanodon. The early scientists who studied fossils frequently had little evidence to go on when trying to reconstruct the whole animal from mere fragments, and Owen battled with other paleontologists about the proper way to construct this dinosaur. A group of Lamarckian-minded anatomists argued that the creature was a much larger, simpler version of our contemporary lizards and should be reconstructed as a belly-crawling reptile. But Owen was fearful that this reconstruction would lend credence to the dangerous idea of evolution, causing people to see this ancient creature as a primitive predecessor to today's more advanced life forms—a validation of Lamarckian progress-based evolution. So Owen's solution was to construct the creature in a more upright quadruped form. Instead of the legs being splayed out to the side, they were brought directly under the body, and the animal was reconstructed more like a mammal (specifically, a rhino) than a reptile. In doing so, he hoped to send a persuasive subtextual message to viewers that there was no progress from earlier to later forms. Owen was *visually* saying, "Look, here's an ancient animal, and it's as complex and advanced as contemporary mammals, so don't get duped by this new progress theory." The irony is that Owen was right (for all the wrong reasons) about the animal's posture.

The fascinating aspect of Owen's dinosaur exhibit is the way that visual presentation, in this case three-dimensional presentation, created a persuasive argument. This is a feature of art that science has been using for ages, but only recently have scholars acknowledged the practice. Archaeologist Stephanie Moser, for example, has called attention to the ongoing use of pictures in the debates over human ancestors. Focusing on the twentieth-century debates about australopithecines, Moser shows how different visual images of protohumans jockeyed for missing-link status and shaped public and professional consciousness about who fell where on the family tree.

In the 1920s archaeologist Raymond Dart began to argue that an australopithecine fossil cranium, which he had discovered in Taungs, South Africa, was in fact a missing link between humans and ancestors who were more apelike. He argued that the teeth and the braincase of the creature showed rudimentary but significant similarities to humans', and he called his discovery *Australopithecus africanus*. For a variety of reasons this was not well received by Dart's professional peers, chiefly because earlier fossil finds had convinced most researchers that Asia would be the proper locus for the missing link. But the discursive arguments were relatively insignificant when compared to the visual arguments. Grafton Elliot Smith, an opponent of Dart's, put together a drawing that showed Dart's *Australopithecus africanus* next to a *Homo erectus* re-creation (fig. 7.12). The drawing is a persuasive visual argument that contends, contra Dart, that *A. africanus* is very primitive indeed, a mere chimpanzee who looks as though he's up on two feet in a precarious and temporary way. All the visual conventions are in place to convince the viewer that Dart's candidate is dim and animalistic: *H. erectus* holds a spear, and *A. africanus* holds a little stone; *H. erectus* wears a loincloth, while *A. africanus* goes naked in all his hairy glory; *H. erectus* is, well, erect (standing, that is), while Dart's boy is in that chimp-after-riding-a-horse stance. Dart responded with images that showed his candidates engaged in all kinds of sophisticated activities, including family groupings that suggested a nuclear domestic structure (remember the grizzlies and titanotheres?). The visual debate raged on for years, with each group attempting to demonstrate the "humanity" of its respective candidate.

Perhaps the most strongly affective image in this ongoing debate was a painting done for *National Geographic* in 1960. Mary Leakey and Louis Leakey had discovered a new australopithecine in Olduvai Gorge, in East Africa, in 1959, and they argued that its tool-making abilities (based upon evidence at the dig) qualified it as the earliest human. To introduce this new australopithecine, which they called

Figure 7.12. Reconstructed australopithecine and *Homo erectus*. Drawing by A. Forestier, under the direction of Prof. G. Elliot Smith, F.R.S. From *The Illustrated London News*, Feb. 14, 1925, p. 239. (THE ILLUSTRATED LONDON NEWS PICTURE LIBRARY)

Zinjanthropus boisei, they had a re-creation portrait painted. Moser describes the portrait as "a close-up of the face of a male figure engaging the eye of the viewer. The species is portrayed as being distinctively human, in the sense that it has a beard and moustache, a fine nose, fine eyebrows, and thin lips." But the truly striking feature is the dignified gaze, the windows-of-the-soul quality of his eyes. Moser points out that the "expression on the face, or the way that 'Zinjanthropus' is depicted looking out at the viewer, suggests that the species is fully self-aware or self-conscious." This new visual convention makes the earlier missing-link candidates look utterly retrograde by comparison. More recent paleontology has again redefined the images of australopithecines, and artworks continue to play persuasive roles in the relevant museums, textbooks, and mass media.

Art and science serve each other in a variety of complex ways: scientific understanding can help artists better express themselves; artistic images can clarify theoretical relationships or otherwise promote comprehension; images are rhetorically persuasive in scientific disputes; and in addition to all these, or perhaps as the very background to all these, there is the affective power of images. The museum uses this affective power as its underlying didactic mechanism. Feelings, emotions, attitudes, and values, all of which can be triggered by images, are employed by museums as the glue that fastens together informational facts.

How is it that all this imagery works so well on us? Probably it's because knowledge is more than propositional and syllogistic. The foundational epistemology of museums is associationism. We started to get into this a little bit in chapter 2, but this is the place for a fuller exploration.

The sensually oriented modern European museums, from the late seventeenth century to the present, assume the empiricist epistemology that John Locke first articulated in his 1689 *Essay Concerning Human Understanding*. Locke attempted to understand the way in which human beings come to know things. "Let us suppose," he says, "the mind to be, as we say, white paper, void of all characters, without any ideas. How comes it to be furnished?" He argued that all knowledge originates in "sensation" and "reflection" (reflection being a kind of internal perception of the operations of our minds, such as doubting, believing, reasoning, etc.). In this way, Locke claimed, everything we know can be explained by sense perception rather than by innate ideas (as previous philosophers had argued). We get all our ideas from experience; we're not born with them. But ideas, even such seemingly simple ones such as "Fast food is unhealthy," can be complex combinations of experiences, and Locke wanted to explain how the combinations get formulated. He claimed that associations are formed in the mind when one perception is frequently coupled (in time or space) with another perception. This leads us to combine separate experiences in our minds, so that when one perception occurs, the other is remembered or experienced with it. When these internal idea combinations correspond with the combinations out in the real world, they're considered "true." But, as Locke pointed out, "whole societies of men" can be worked into "universal perverseness" because unrelated experiences "of no alliance to one another, are, by education, custom, and the constant din of their [political] party, so coupled in their minds, that they always appear there together" and become confused as one idea.

David Hume continued and refined Locke's associationism throughout the eighteenth century, in his 1739 *A Treatise of Human Nature* and his 1758 *An Enquiry Concerning the Human Understanding*. Hume went so far as to argue that even our understanding of causality itself (e.g., that fire causes heat) is nothing more than a constant conjunction or recurrent coupling of perceptions (sight of flame, tactile sensation of heat). Perceiving a causal connection between things is really just acknowledging the most routine association of impressions. So, according to Hume, the very best of human knowledge (science, as an example) and the very worst of human knowledge (prejudice) originate in fundamentally the same process. Good learning, then, is a matter of noticing and reinforcing the right sort of associations, meaning the associations that accord with regularities in nature. Exactly *how* you distinguish whether your associations accord with nature (since you only have other associations to check them against) has been giving philosophers migraines for years, but it need not concern us here. What's important to see is that knowledge, according to associationism, is a matter of coupling sensory experiences.

The psychological school of behaviorism was just a version of associationism that tried to move the significant couplings from inside the head to the outside behaviors. So too, today's artificial-intelligence research on neural-network computers that can "learn" is based on an associationist epistemology. But this basic view of learning actually goes back to the medievals.

In medieval Europe, where reading and writing were uncommon skills, it was important to develop a good memory. In a literate culture, one can store information in places other than one's mind (though one could joke that doing so atrophies the brain so much that people like me have to write new phone numbers and such on their hands). When medievals needed to memorize a lot of information (legend has it that Avicenna memorized the entire works of Aristotle), they revitalized a classic method of mnemonics. One could store and recall huge amounts of data by using something called a "memory theater." Using your imagination, you would mentally construct a building in great detail (entranceways, corridors, different-sized rooms, etc., all conceived very distinctly). Next you would imagine the proper lighting arrangements (and this was important, because if the lighting was too intense, it would blind your memory; if too dark, it would muddle things). After mentally constructing your memory theater in this way, you would sequentially and intuitively place the materials to be memorized inside the rooms. Remembering something complex, then, would involve a mental tour through your theater.

Placing your memory material (say, Aristotle's many ideas) into these rooms, however, had to be done in a special way. The medieval mnemonic bible, the *Ad Herrenium*, recommended that ideas could be best remembered if they were imagined as strong images. In this way the information would be distinct and obvious. The images that you use should be unusual, ornamental, funny, or bloody. So if you wanted to remember Aristotle's theory of cathartic tragedy, let's say, you might imagine the toilet room, where . . . well, you get the picture.

Medieval culture was a very visual, image-based culture. The medievals understood the role that images can play in our intellectual lives, and they employed and exploited imagery to teach and to learn. Church frescoes, for example, could serve as condensed sermons about moral activity. A set of well-done images could communicate complex cause-effect relationships to worshipers. When you're tempted, for example, to covet your neighbor's wife, you might not remember the priest's lengthy homily, but you might remember the Bosch-like fresco of that guy with the arrow up his ass and the demons around his throat.

Associations, especially using visuals, can be very powerful in the acquisition, contemplation, and recall of knowledge. The philosopher of science Daniel Dennett has invented a nice term for this nonsyllogistic kind of thinking: "intuition pumps." We can fix upon an image, a metaphor, or an expression and use it as a platform for further thought and understanding. Dennett describes these with regard to the philosophical tradition, but his claims also apply nicely to the museological intuition pumps (such as the roulette wheel and the dioramas generally).

> If you look at the history of philosophy, you see that all the great and influential stuff has been technically full of holes but

> utterly memorable and vivid. They are what I call "intuition pumps"—lovely thought experiments. Like Plato's cave, and Descartes's evil demon, and Hobbes's vision of the state of nature . . . I don't know of any philosopher who thinks any one of those is a logically sound argument for anything. But they're wonderful imagination grabbers, jungle gyms for the imagination. They structure the way you think about a problem.

Most museum display developers feel very successful if their visitors come away from the exhibits with an intuition pump or two. No one expects the patron to retain large amounts of text-based factual material or theoretical minutiae. But if some conceptually rich images get lodged in people's brains during their visit (e.g., roulette wheels, chunks of unrecycled garbage, upright iguanodons, cladogram walking paths, etc.), then these might help to structure the way they think about issues and problems in the future.

The British collector Henry Pitt-Rivers (1827–1900), whose original display technique is still preserved in the Oxford University Museum, argued that the proper placement of his ethnological specimens could have a profound impact on the political convictions of his visitors. Pitt-Rivers laid out his ethnological collections in typological arrangements that stressed the very gradual changes and underlying continuity of human tool designs. He explained the political associations that he hoped his displays would create in the minds of working-class visitors: "Anything which tends to impress the mind with the slow growth and stability of human institutions and industries, and their dependence upon antiquity, must, I think, contribute to check revolutionary ideas, and . . . inculcate conservative principles, which are needed at the present time, if the civilisation that we enjoy is to be maintained and to be permitted to develop itself."

Today's curators are less ambitious about the impact of their specific displays, but they agree about the basic epistemology. Minda Borun, a curator at the Franklin Institute Science Museum in Philadelphia, points out that many adults (not just children) hold "naive notions" about the natural world, and museums are challenged to restructure or retrain these cognitive associations: "Unless they have experiences that cause them to become aware of the flaws or limitations in their early explanations, people's naive notions persist."

As an example, Borun explains that the Franklin Institute discovered, through front-end surveys, that many adults naively associate gravity with air pressure, and when they were asked whether a ball would fall in a vacuum, they predicted that the ball would float. To rid visitors of this incorrect association and replace it with correct understanding, curators attached a glass tube to an air pump and placed a Ping-Pong ball inside the tube. When the visitor rotates the tube, the ball falls from one end to the other. Then the air is pumped out of the tube and, on further rotations, the visitor discovers that the ball still falls as before. Borun argues that exhibit planners must first discover the patterns of incorrect associations that are present in the public and then work toward creating corrective associations.

When I interviewed the Field Museum's senior exhibit developer Eric Gyllenhaal, he explained that the associations exhibit developers want to engender in visitors are the ones that will help them understand and construct scientific knowledge after they leave the museum. The epistemology may be associationist, but the developers do not treat the visitors as passive recipients of data. Eric explained that "if you have an understanding that education is simply taking the information that we have and giving it to people, and they'll just accept it as we give it to them, then, it will inevitably lead to more and more emphasis on entertainment; because you're thinking of visitors as passive receivers. But so too, on the other hand, the curators who just want to put more and more content in, are another manifestation of that more passive model."

Alternatively, Gyllenhaal believes, we can understand knowledge as being something that museum visitors have to construct for themselves, and therefore curators want them to understand how that knowledge is constructed by scientists. In that way, visitors can continue to construct new knowledge as it comes along from newspaper stories, television, and whatever else they're exposed to.

The museum itself, then, can actually model scientific thinking by walking the visitors through the critical-thinking skills that researchers employ. Eric points out that this is the new rhetoric that museums use in order to request and justify public money. The institutions now argue for funding on the grounds that they are helping visitors to build their own knowledge and appreciate what they themselves actually know and can know, as well as the process of how they know things.

This recent attempt to expose more of the scientific process has led to some interesting museological trends in the last few years. For one thing, at some natural history museums the exhibits are deliberately more humorous. Some might dismiss this new trend as yet another edutainment move that simply dumbs down content and seeks to create amusing diversions. But much more is happening when dioramas and displays express a little wit. Perhaps the most obvious power of exhibit humor is its ability to associate positive moods (it doesn't get more positive than laughter) with learning. Humor directly addresses Bloom's affective domain of education.

Some museums have more of a sense of humor than others. Moreover, some museums engage in different kinds of humor. The French Grande Galerie de l'évolution, for example, contains very little mirth, but when it does express wit (for example, when it hung a series of misshapen specimens from the ceiling with no explanation), it tends to be very dry, clever, and inward-focused. Britain's Museum of Natural History, like the Field Museum, employs many game-show-type exhibits; this exhibit strategy kills two birds with one stone because it's both funny and interactive. Developers at South Kensington also carefully punctuate dense informational sections with intervals of comic relief. For example, after a display on complex paleontology techniques, a wall of cartoons follows that illustrates funny theories about how dinosaurs became extinct (boredom, mass suicide, smoking, and so on). There was also a display case that was crammed full of hundreds of brightly colored plastic dinosaur toys. When I was there, kids were mobbed around the case, laughing, smudging the Plexiglas, and shrieking with excitement.

The American Museum of Natural History, in New York, is generally very sober. The display philosophy has a certain staid Victorian quality; the nostalgia is not without its charm, but humor seems contrary. The one exception to this AMNH rule is a knee-slapping display of a human family lounging around in their natural living-room environment (fig. 7.13). But the humans are all skeletons, engaged in activities such as reading and watching television. One of the bony specimens reclines in a comfy chair, reading *The Human Odyssey*. Another sits watching TV while an ossified dog poses beneath a skeleton version of Grant Wood's "American Gothic." Even the wallpaper designs consisted of free-floating femurs and fibulas. The display self-consciously, and unapologetically, revels in bones; it's a museum exhibit that pokes fun at museum exhibits. This humorous breather is so unlike the rest of the AMNH that one wonders if they sacked the development team right after it was unveiled.

Nowhere else in my travels did I see a museum that so consistently goes for the funny bone as Chicago's Field Museum. Large segments of the Field's "Life over Time" exhibit seem to have been designed by Monty Python's Flying Circus. One of the displays informs the visitor about the evolution of mammalian teeth, and it does so by re-creating a dentist's office, complete with a cow skeleton sitting in the dentist's chair awaiting a checkup. Behind the skeleton, the cow's Holstein skin hangs on an office coatrack (fig. 7.14). In another exhibit about animal morphology, different taxonomic categories are listed on a large display board, and small animal models are glued next to their class names. The display is quite sober until you reach the tiny "Mammals" sign, where a two-inch toy woman (another cereal-box specimen?) waves out at you; accompanying a label that reads "*Homo sapiens*, model (not life size)." In some places, such as the *Homo sapiens* "birthday party" (see fig. 6.7 in the previous chapter), the humor disarms the visitor's natural hostilities. Humor can be a rhetorical device that gets people to let their guard down a bit, especially when it comes to certain controversial topics such as human evolution.

Humor can also help generate important associations, both by setting positive moods and by standing out for memory recall. The Field's "Life over Time," for example, employed a video-monitor edutainment technique that really stuck in people's minds. In addition to the funny Godzilla video that played in the dinosaur hall, the entire exhibit was peppered with video loops of famous network news anchors giving humorous (but accurate) news reports about weather conditions during the Cretaceous, evolutionary trends during the Devonian period, and so on. Eric Gyllenhaal told me that these video segments were cited in visitor surveys as one of the exhibit's most memorable features, even as much as six months after their Field visit. Even visitors who didn't particularly like the videos still remembered them and wanted to talk about them. Visitors

Figure 7.13. A humorous display of a skeletal human family, relaxing in their natural living-room environment. Foreground figure reads a book in a lounge chair, while a baseball game plays on the television in the background. American Museum of Natural History, New York.

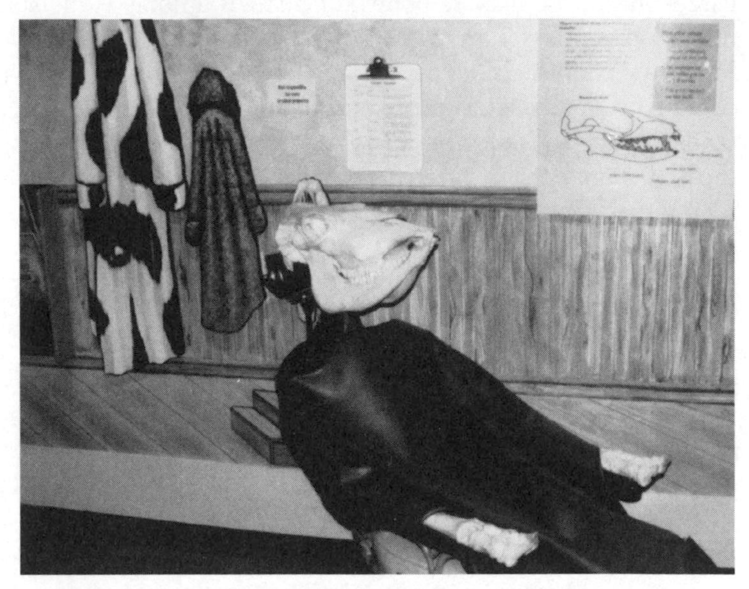

Figure 7.14. Humorous display on mammalian teeth in the Field Museum's "Life over Time" evolution exhibit. A cow skeleton sits in a dentist's chair, awaiting a checkup. Notice the Holstein skin hanging on the coatrack in the background.

grasped some sense of the interconnection of geological, meteorological, and biological change from the newscasts.

The self-mocking irony of some current museum exhibits is a complicated phenomenon. On one hand, it reflects something about today's mass-media-savvy culture. Most of us have been raised on advertising and are, or at least pretend to be, suspicious of sincerity. Perhaps because of this immersion in the mass media, or perhaps as a result of their loss of innocence about politics, baby boomers and Generation X'ers may be especially cynical and sarcastic about institutional displays generally. When the propaganda of consumerism saturates so much of one's popular culture, one tends to approach other images, presentations, and display communications with mistrust and trepidation. The fact that natural history exhibits are now made by and for this generation makes some display irony inevitable. But some self-mocking commentaries can also actually help developers communicate more about the nature and process of science itself.

Think about the Musée de l'homme's funny display of human evolution, with the visitor's face at the ultimate spot in the sequence of skulls; or the AMNH's skeletal family, lounging in their TV-room habitat; or the Grande Galerie's homage to the quirky curiosity cabinets of yore. These wry museological self-references can actually help visitors feel closer to the biological sciences. This kind of display irony, which laughs at its own historical and contemporary conventions, exposes the human side of biology to the visitor. When an institution laughs at itself, it becomes instantly more approachable because it temporarily relinquishes its authority status—even if only momentarily. Institutional authority, with its completely earnest and sincere conventions, is alienating for today's museum-goers. A museum's majestic architectural spaces can have an affective impact on the visitor, inspiring awe and reverence (museum as temple), and this can be important because it can lead to meditation, inquiry, and wonder (fig. 7.15). But self-deprecating jokes are great, too. They can lead visitors to see past the institutions to the real authorities of science, namely, reason and experiment—two things that visitors are not alienated from. Humor in museums can give visitors a glimpse of the fallible nature of science itself, and this ensures that a healthy skepticism toward sacrosanct truths will be maintained.

Museums are increasingly acknowledging the fallible character of scientific knowledge. Until recently, museum rhetoric and science education generally have tended to present scientific truth as a fait accompli. The controversies surrounding most interesting scientific ideas have been hushed up or downplayed in order to present a unified front to the supposedly confused layperson. Museum educators, for example, have been loath to present exhibits about the internal disputes regarding evolutionary mechanisms and tempos, for fear that these scientific disagreements will fuel the unrelated cultural fires of antievolutionism. In some cases they're right to be afraid, because scientists who have been researching nonselection types of

Figure 7.15. Image from the Field Museum's great room. Note the "majestic" and "sublime" qualities of the classical architecture and natural *Apatosaurus* skeleton.

transformation occasionally appear, fully distorted, as citations in creationist literature. Nonetheless, it does no good for science to pretend to be something that it is not: perfect.

Absolute certainty is an intense and impossible demand that our emotions make on our intellect. There is something comforting and soothing about reliable, absolute, and unchanging truths. Unfortunately, you can count all of them on one hand. Until recently we have tended to see science and society as two different realms, and science is often interpreted as transcending the human foibles that plague the rest of society. It was assumed that the neutral scientific method (the hypothetico-deductive model) could completely expunge all the subjectivity of interpretation from the scientific process, leaving a clear vista of truth that is free of controversy. This overly sanguine representation of science has led to science worship and science hatred, but not science understanding. Many museums are now trying to exhibit the fact that scientific theories change over time; they get revised, they get thrown out, they get forgotten. Traditionally speaking, the scientific community has feared the possible relativism that might arise if the lay audience was exposed to the "real thing" and has worried that if Jane Q. Public realizes that science often succeeds without following the classical scientific method, then she might lose faith. Chances are, however, that she might begin to appreciate how science, like every other human enterprise, is imperfect, but still a very powerful (indeed, the most powerful) tool for understanding nature. Instead, the science community, frequently influenced by the mass media's need to soundbyte everything, projects itself as a superhuman, unerring, fact-crunching machine. The wider culture is shown the products of science but not the process. Then, when science makes mistakes or changes its mind, it looks manipulative and hypocritical, leading to increasing science phobia.

All the imagery of science as authoritarian, however, is utterly contrary to the genuinely empirical spirit of science. Many natural history museums have been changing their text plaques and labels to use language such as "This animal *probably* did that," or "This anatomical structure *suggests* this function," or just plain "We *don't know* why this did that." To scientific dogmatism, these admissions of uncertainty are like heresies. And in addition to new language, many museums have begun to show parts of the backstage work in the front-stage exhibit spaces. Curator Sally Love, at the National Museum of Natural History, claims that earlier curators only exhibited the end products of the behind-the-scenes scientific research. Then researchers and curators were disappointed when the public wasn't getting it, and some curators simply chalked it up to visitor ignorance. But, she points out, the real problem was that visitors, exposed only to end products, were being excluded from the learning process. Besides, she says, "people are curious about what they can't see. Closed doors and solid walls seem to hide secrets about how exhibits and museums work." One way to give visitors access is to punch windows from the exhibit spaces through to the backstage work spaces. This is a growing trend, and one can witness a great deal of it at the South Kensington museum and the Field Museum.

At the Natural History Museum in London, an innovative new project is currently under way. Over the next five years they are constructing the Darwin Centre, a major capital investment that, according to their proposal, will "throw open the doors of the Museum to the public in completely new ways, by revealing for the first time not only the amazing organisms in its collections, but also the cutting-edge scientific research they support." The first phase of the project will center around zoological collections, and it promises to incorporate "unprecedented public access to scientific laboratories and collection stores."

In addition to increasing direct backstage access, museums are currently re-creating mock backstage representations within the traditional front-stage display spaces. At the Field Museum, for example, there was a corner exhibit of a re-created paleontology office. It was a mock work desk with small dental tools for fossil cleaning, a stack of academic books, and an answering machine that looped an excited (but now warbly) voice saying, "Jane, you've got to come down to this dig site, you'll never believe this amazing discovery!" When I interviewed Eric Gyllenhaal, I asked him about this particular exhibit, and whether the Field was consciously trying to show visitors more backstage material. He confirmed that the paleontology office display was indeed an explicit attempt to show visitors what the hidden researchers were doing with their days.

"We almost did a whole exhibit," he said, "on 'behind the scenes at the museum.' That's one thing I was working on between 'Life over Time' and 'Life Underground.' I think natural history museums in particular are concerned that the public doesn't understand what they really are, and we think that this fact has implications for our ability to get funding as institutions that do research. And so places like the Smithsonian's natural history museum in particular, but really, natural history museums all across the country, are trying to educate their public about what goes on behind the scenes in terms of research. So that particular interest has been going on here at the Field Museum at least since the early nineteen eighties. Several times, in fact, I've tried to get major exhibits out about it, but I've had to settle for smaller behind-the-scenes sections in other exhibits."

The mock office exhibit, Eric explained, is just another manifestation of exhibit planners' general interest in displaying the culture and process of science. Scientists want to have nonscientists understand science the way they understand it. And you can see this same tendency manifested in the parts of the exhibits where they talk about the things they don't know yet, such as the uncertainty about whether dinosaurs were warm-blooded or cold-blooded. Of course, natural history museums are taking a big risk here as well, because opening the doors on the back-room operations sort of assumes that the public will appreciate what's going on back there. Scientists tend to think that if the public only knew what they did, they'd be appreciated. It's unclear yet whether this is true. Another risk exhibitors are taking by opening up the research doors is that local government officials will suddenly realize that there's all this research going on behind the scenes and think that the museums are somehow cheating the public by taking all this money and putting it into research instead of exhibits. Eric explained that the Field needs funding from philanthropic organizations, corporations, and the government just to survive. It costs about $30 million a year to operate the museum ($9 million of which goes to fund the research scientists), and the museum takes in only about $4 million a year in ticket admissions and memberships. But if exhibitors can get the public to accept and appreciate the research, then this will be important when it comes time to justify the budget.

At the same time that current exhibit developers are attempting to reveal the backstage culture of museums, that culture is itself undergoing significant change. Field Museum curator John Wagner explains that recent years have seen increased institutional pressure on curators to become nationally and internationally prominent characters. It is not uncommon now to find stories and blurbs featuring high-profile paleontologists such as Dr. John Flynn (of the Field Museum and the American Museum of Natural History) and Dr. Paul Sereno (of the University of Chicago and the Field Museum) in popular magazines and television news programs. Today's museums want real-life Indiana Joneses, or at least media-savvy charismatic figures who can stay visible.

"Nowadays," Wagner says, "if you want to survive in the museums, you have to be very good at getting grant money; you have to publish like a fiend; you need to be involved in public programs; you should have an adjunct professorship at one of the universities, and you should be able to speak fluently and enthusiastically to board members and visiting dignitaries. In short, you need to be a successful political animal these days to survive in the museum. And, of course, when that charisma is coupled with real subject-matter expertise, it's a wonderful combination. But often it is more flash than substance.

"Recently, museums are putting such an emphasis on high-profile notoriety that the collections themselves are being neglected by the new curators, who spend all their time promoting themselves. The sad fact is that many quieter people, who put in years of good work at the Field Museum, have recently lost their jobs to more dynamic but less educated competitors. The nature of the work, hunched over tiny bugs or fossils in a hidden-away cubicle, for example, traditionally drew introverts to the curator and staff jobs. And the museum nurtured them. It's a quirky person that is willing to sit at their little station and study minute details for years, and get excited about it; and poke and prod at their specimens until they crunch out a little bit of truth about them. And that kind of work does something to you. If you can come through that process and possibly find a mate and raise a family and be a relatively normal member of society, then that is absolutely amazing.

"The tragic thing," Wagner continued, "is that the current trend is for museum trustees and administrators to ignore the internal, albeit quirky, talent when staffing positions of power and go outside for M.B.A.'s who frequently don't know anything about the nuances of the subject matter. The recent national profusion of graduate programs in management, for example, have produced lots of managers who don't know what they are managing. And in many cases, the quirky but qualified people lose their jobs altogether. I guess it's museum, red in tooth and claw."

The direction in which natural history museums will be going in the next millennium is unclear. Many of the larger institutions are undergoing significant transitions. Whatever happens down the road, it is clear at least that a major player in the immediate transition will be "Sue." Sue will be front-stage, back-stage, side-stage, you name it. As I write this, Sue is the jewel in the Field Museum's public relations crown, because she single-handedly does all the important edutainment work that Eric mentioned above. She represents the wonders of nature, by being the biggest and most complete *Tyrannosaurus rex* ever excavated (forty-five feet long). She represents the culture and process of science, because the Field is having her bones painstakingly cleaned in a front-stage glass lab called the

McDonald's Fossil Preparation Laboratory (fig. 7.16). She ensures that the Field will continue to be a world-class natural history institution, and she sets a precedent and a model for the international museum community. Sue was formally unveiled in 2000.

Sue, named after her discoverer, Susan Henrickson, was at the center of a huge legal battle that started after her discovery in 1990 in South Dakota. The fossil *T. rex* was discovered on the property of Maurice Williams, a Sioux Indian whose land was held in trust by the federal government. A fossil hunter named Peter Larson paid Williams a mere $5,000 for Sue, but when the feds got wind of it, they confiscated the bones. The government, being the trustee of the property, felt justified in reclaiming the fossil because it had never been asked for permission to remove the bones. The government decided to auction Sue and pay Maurice Williams the proceeds.

In October 1997 Sue was auctioned at Sotheby's in New York. The bidding was intense, to say the least. Sotheby's had expected to get at least $1 million for the fossil, but the competitive bidding drove the final price up to $8 million. In order to pay out such an enormous amount, the Field formed a partnership with the McDonald's Corporation and Walt Disney World. Naturally, McDonald's and Disney get something out of the deal. In addition to the fossil preparation laboratory, McDonald's gets two life-sized replicas of Sue that will travel nationally and internationally. Walt Disney World's Dinoland USA, which is sponsored by McDonald's, will also receive a life-sized replica. McDonald's chairman Jack Greenberg explained after the auction that Sue was "McDonald's gift to the world for the millennium." Look for little plastic toy Sues in kids' Happy Meals soon (fig. 7.17).

To my mind, Sue represents the best and the worst of edutainment. Sue is the very embodiment of good museum drama. She is ancient in that way that gives visitors a brush with human fragility and finitude. She is dangerous and imposing (her teeth are twelve inches long) in the way that simultaneously repels and attracts visitors. She's rare and strange, and she kindles a bonfire of imagination and research. Her bones tell many tales that researchers and visitors will be trying to decode for years to come. For example, the fact that she broke her leg (usually instant death for a predator) and that it was partially healed suggests that *T. rex* may have had continuous mates (a male that brought food, in this case). Sue will be a prism for many such issues of ethology, physiology, morphology, and so on. Besides all that, she's just plain cool.

Sue is also an excellent interface point for art and science, an example of the confederacy of these two domains that we've been

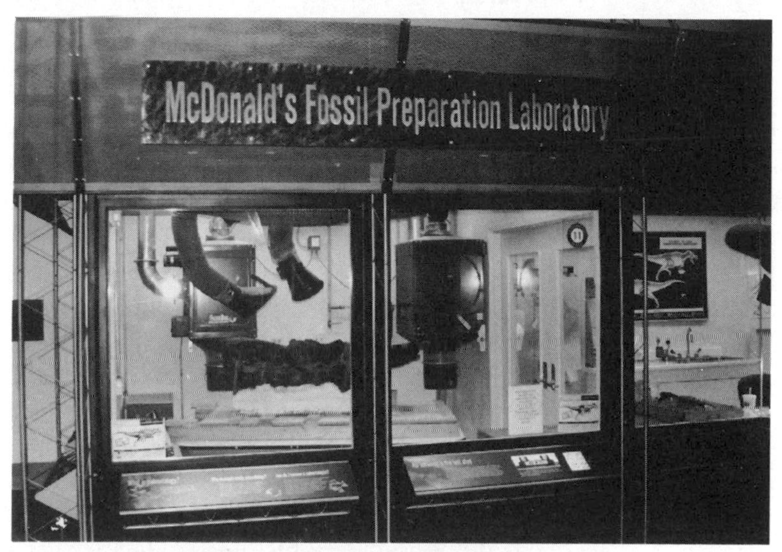

Figure 7.16. McDonald's Fossil Preparation Laboratory, located in Chicago's Field Museum. Paleontologists clean the bones of the *T. rex* nicknamed "Sue" in a glass enclosure so that museum visitors can view the process.

Figure 7.17. Sign advertises corporate restaurant at Chicago's Field Museum.

exploring throughout this chapter. The associate director of Sue's exhibition is named Rich Faron, and his formal training is in sculpture and art history. He explains that he is "exploring ways to create theater around the skeleton." Drama is, of course, a feature of museum knowledge that goes back to the earliest dioramas and habitat mounts, but Faron clarifies this by saying, "I'm the director, and the curators are the playwrights. You have to take an artifact and stage it, give it energy and motion and dynamics. You enter an exhibit like you would a theater, ideally. There should be anticipation and an unfolding for the visitor, a story that should build toward some sort of conclusion—an aha! moment. If you get people to that point, you are teaching them something." Sue can, and will, have tremendous affective power for museum visitors in the future.

The unsettling feature of the Sue story is the obvious one. McDonald's must be figuring something like "If families support learning (about things such as *T. rex*), then they'll support McDonald's." Given what we know about associationist knowledge, we cannot fault advertising for adopting the impression-coupling strategy. But megacorporations are now infiltrating museums at their very core. One cannot adopt a naive posture here. Admissions alone leave the Field Museum, for example, $26 million short for annual operating costs, and this suggests that museums of the future will always have some intimate relationship with corporate funding. It is a beneficial venture for both parties. But when big business has such a powerful hand in the acquisition and display of collections, one can't help but be concerned.

All of the contemporary museums that I toured (including the Grande Galerie and the Musée de l'homme in Paris, the Museum of Natural History in South Kensington, the Oxford Museum of Natural History, the American Museum of Natural History, and the Field Museum) printed the names of corporate and private sponsors on plaques in the exhibits and in gift-shop brochures. Most museums have struck a deal with their corporate sponsors, wherein they agree to print the business's name but *not* the corporate logo. I suspect that very soon, however, one of the corporate sponsors who are putting up millions of dollars for specimens will tell a museum that the logo comes attached to the money. And when museums are so dependent on their corporate investors, they will have no leverage by which to refuse the commercial demands. Money always has strings attached, and museums will have to be careful in the coming millennium.

When corporations become deeply rooted in educational institutions such as museums, they can prejudice the information that gets presented. For example, all over the country there are science and industry exhibits about petroleum sponsored by various oil companies. Guess how many of them devote exhibit space to the problems of air pollution.

In addition, the more dependent that museums become on fast-food corporations, oil companies, movie tie-ins, and bloated marketing strategies, the less likely they will be to take risks. Just as in the case of Hollywood movie studios, if museums move increasingly toward blockbusters, the huge investments these require will inevitably force them to move increasingly toward formula exhibits. Museums have to be careful to keep some autonomy from big business, but also some autonomy from popular opinion.

This last point may seem at first like antidemocratic elitism. The growing trend of front-end surveys and predevelopment marketing research may seem more in tune with giving the people what they want, but that's not the best criterion for developing a good education program. Remember the Harris loan program's suitcase dioramas at the Field Museum (chapter 1). The educators there were very nervous about creating evolution dioramas, and their nervousness is understandable, because the cases go directly to grade schools, the precise loci of tension between Darwinians and creationists. But if all this front-end surveying tries to assess what the public wants to see exhibited, then even large-scale evolution exhibits such as "Life over Time" might start to vanish, and curators and developers may become too afraid to communicate controversial truths. Political prejudices in the public sphere, for example, might be reflected in the polls and surveys that increasingly dictate the most important topics and trends for museology. Educators should challenge us to move beyond ourselves, not just reflect our own ideas back at us. We, the public, need to have faith in our educators (in this case, museologists), and they need to have faith in themselves to set educational agendas without constantly checking with us to see if we approve. When I meet with my undergraduate students every new term, I give them an informal front-end evaluation to find out what things they are interested in and excited about. I try to work in some of these things during my courses, because this is how you connect the student to the more remote information or the more unfamiliar material. But if I tried to base my class on the topics that emerged out of the front-end evaluation, then my entire course would focus on which pop music bands "rule" and why dating is so frustrating. This would make for a pretty pathetic course on the history of science.

My suspicion is that museums of the next millennium will be able to navigate between the Scylla of big business and the Charybdis of popular taste, but we have to remember that the museums we all grew up on are still only infant institutions. They have only recently, relatively speaking, made the transition from private collection to public resource, and they continue to define and redefine themselves every decade. The coming age, with its Internet and virtual-reality technologies, may see the birth of important cybercollections or cybercabinets. Museums may digitize themselves completely in the coming century, but this will only continue the low-tech virtual-reality tradition of taxidermy. It is hard to say what the implications will be when people immerse themselves further into artificial and synthetic means of encountering nature. In a sense, we already live in this strange world, where just by sitting in front of your TV you can know more about elephants or lions than Buffon ever knew. Soon you'll be able to put on a virtual-reality glove and real-time video goggles and actually pet a lion in Kenya from your living room in New Jersey. Of course, in this approaching world, you will probably never meet an actual lion. But perhaps the art of collecting, with its penchant for the genuine and the rare, will be even more engrossing in such a world.

Traditionally human beings have felt quite comfortable drawing a distinction between art and nature. While Aristotle claimed that the term *nature* should be ascribed to those things that "have their source of motion or activity within themselves," and that art is only a product of external motions (human artifice), the rest of us have also had a vague and intuitive sense that this art/nature dichotomy is an accurate way to carve up the world. But this once intuitively clear distinction has now been blurred beyond any recovery. Our folk taxonomy is threatened daily, because humans can now take their prosthetic experiments to the inner workings—the very blueprints—of organic life. Indeed, the word *biotechnology*, which would have sounded like a contradiction in terms to earlier eras, has become completely commonplace. The idea that nature and human artifice are autonomous categories is simply untenable in this age of cloning, genetic engineering, and neuropharmacology. And yet, in a less obvious way, the distinction has *never* been entirely tenable, because nature has *always* been a human construction, a product of human artifice. Touring these museums demonstrates that nature itself is conceptualized, presented, and constructed in different ways by different eras and cultures. Frequently this packaging of nature goes unnoticed because it is accomplished quietly by curators behind the scenes. And it is occasionally accomplished unconsciously by the paradigm structures that happen to be in place.

I started this book trying to figure out how to "stuff" a human being for display, an interest that, frankly, I'm rather embarrassed about. Still, it increasingly led me to go behind the scenes of nature collecting. But when I got behind the scenes, I found that there was more than just specimen preparation labs and viscera. My journey was also a conceptual venture behind the scenes, with equal shares of animal corpses and abstract ideas. The cultures of collecting, I discovered, are always as curious as the collections themselves.

NOTES AND FURTHER READING

INTRODUCTION

Some excellent books on collecting and museology have influenced my work. Some of these books include: Krzyztof Pomian, *Collectionneurs, amateurs et curieux* (Paris: Editions Gallimard, 1987); Michael M. Ames, *Cannibal Tours and Glass Boxes* (Vancouver: University of British Columbia Press, 1992); Victoria D. Alexander, *Museums and Money: The Impact of Funding on Exhibitions, Scholarships, and Management* (Bloomington and Indianapolis: Indiana University Press, 1996); Douglas Preston, *Dinosaurs in the Attic: An Excursion into the American Museum of Natural History* (New York: St. Martin's Press, 1986); *The Cultures of Collecting*, edited by John Elsner and Roger Cardinal (Cambridge, Mass.: Harvard University Press, 1994); *Visual Display: Culture Beyond Appearances*, edited by Lynne Cooke and Peter Wollen (Seattle: Bay Press, 1995); Eilean Hooper-Greenhill, *Museums and the Shaping of Knowledge* (London and New York: Routledge, 1992); Paula Findlen, *Possessing Nature* (Berkeley: University of California Press, 1994); *Exhibiting Cultures: The Poetics and Politics of Museum Display*, edited by Ivan Karp, Steven D. Lavine, and Christine Mullen Kreamer (Washington, D.C., and London: Smithsonian Institution Press); Susan M. Pearce, *Museums, Objects and Collections: A Cultural Study* (Washington, D.C.: Smithsonian Institution Press, 1992); *Objects of Knowledge*, edited by Susan Pearce (London and Atlantic Highlands: Athlone Press, 1990).

CHAPTER I

More detail regarding Foma and Peter's other collecting habits can be found in Oleg Neverov, "His Majesty's Cabinet and Peter I's *Kunstkammer*," translated by Gertrud Seidmann, in *The Origins of Museums*, edited by Oliver Impey and Arthur MacGregor (Oxford: Clarendon Press, 1985).

The facts concerning the American Museum of Natural History's acquisition of Inuit "specimen" Qisuk and his son Minik can be found in Charles Trueheart's article "The Eskimos Finally Go Home," *Washington Post*, July 6, 1993, C1.

Further information about Sigmund Freud's psychology of collecting can be found in his 1908 "Character and Anal Erotism," in *The Standard Edition of the Complete Psychological Works of Sigmund Freud*, edited and translated by James Strachey, vol. 9 (London: The

Hogarth Press, 1953–74). Also see John Forrester, "'Mille e tre': Freud and Collecting," in *The Cultures of Collecting*, edited by J. Elsner and R. Cardinal (Cambridge, Mass.: Harvard University Press, 1994). Also see Lynn Gamwell's "A Collector Analyses Collecting: Sigmund Freud on the Passion to Possess," in *Excavations and Their Objects: Freud's Collection of Antiquity*, edited by Stephen Barker (Albany: State University of New York Press, 1996). The quoted passage by Jean Baudrillard is taken from his "The System of Collecting," in *The Cultures of Collecting*, edited by J. Elsner and R. Cardinal (Cambridge, Mass.: Harvard University Press, 1994).

The information regarding Darwin's study of taxidermy is taken from R.B. Freeman, "Darwin's Negro Bird-Stuffer," in *Notes and Records of the Royal Society of London* 33 (1978). For a better understanding of the state of taxidermy during the late nineteenth and early twentieth centuries, see John Rowley, *Taxidermy and Museum Exhibition* (New York: D. Appleton and Co., 1925). Jonas Brothers' taxidermy catalog of 1965 is titled "Game Trails" (Denver: Jonas Bros., Inc.). The quotes from Nadine Roberts come from her no-frills text, *The Complete Handbook of Taxidermy* (Blue Ridge Summit, Penn.: Tab Books, 1979).

For more information on the ecological and systematics research under way in the Amazonian rain forest, see the special report "Fragments of the Forest" in *Natural History* 107, 6 (1998): 34–52. For a detailed discussion of the use of DNA hybridization in current systematics, see Charles Sibley and Jon Ahlquist, *Phylogeny and Classification of Birds: A Study in Molecular Evolution* (New Haven: Yale University Press, 1991); also see their influential article "DNA Hybridization Evidence of Hominoid Phylogeny," *Journal of Molecular Evolution* 26 (1987). An excellent discussion of the relationship between homology research and molecular phylogenetics can be found in Roger Lewin, *Patterns in Evolution: The New Molecular View* (New York: Scientific American Library, 1997). To better understand the evolution of the concept of homology, see Toby Appel, *The Cuvier-Geoffroy Debate: French Biology in the Decades Before Darwin* (New York: Oxford University Press, 1987).

The quote from Carl Sagan is taken from his book *The Demon-Haunted World: Science as a Candle in the Dark* (New York: Ballantine Books, 1996). The passage quoting Stephen Jay Gould is taken from his *Dinosaur in a Haystack* (New York: Harmony Books, 1995). The quote by Stephen Toulmin is taken from his *Cosmopolis: The Hidden Agenda of Modernity* (Chicago: University of Chicago Press, 1990).

Further information regarding the early days of the Harris loan program (the suitcase dioramas) can be found in *Field Museum and the Child* (Chicago: Field Museum of Natural History, 1928). Quotes of Albert Eide Parr are taken from *Three Addresses Delivered at the Meeting Commemorating the Fiftieth Anniversary of the Field Museum of Natural History, Sept. 1943* (Chicago: Field Museum of Natural History, 1943).

For further analysis of Hornaday's taxidermy, see Sally Love, "Curators as Agents of Change," in *Exhibiting Dilemmas: Issues of Representation at the Smithsonian*, edited by Amy Henderson and Adrienne L. Kaeppler (Washington, D.C., and London: Smithsonian Institution Press, 1997); and William T. Hornaday, *Our Vanishing Wild Life* (New York: Charles Scribner's Sons, 1913).

The quote by Peter Crane, director of the Field Museum, is from a special report to *Museum News* entitled "Toward a Natural History Museum for the 21st Century," November/December 1997.

The quote by Robert Hooke is taken from Bert S. Hall's article "The Didactic and the Elegant: Some Thoughts on Scientific and Technological Illustrations in the Middle Ages and Renaissance," in *Picturing Knowledge*, edited by Brain Baigrie (Toronto, Buffalo, and London: University of Toronto Press, 1996).

CHAPTER 2

The quoted material regarding the front-stage/backstage distinction in funeral homes is taken from an excellent sourcebook of dramaturgical sociology entitled *Life as Theater*, edited by D. Brissett and C. Edgley (New York: Aldine de Gruyter, 1990).

The quotes of Jack Kevorkian are taken from his *The Story of Dissection* (New York: Philosophical Library, 1959). For a very comprehensive explanation of the little-understood practices and processes surrounding dead bodies, see Kenneth V. Iserson, *Death to Dust: What Happens to Dead Bodies?* (Tucson, Ariz.: Galen Press, Ltd., 1994). I have included some passages from Jessica Mitford's humorous and insightful *The American Way of Death Revisited* (New York: Alfred A. Knopf, 1998). Also see Clarence Strub and L.G. Fredrick, *The Principles and Practice of Embalming* (Dallas: L.G. Fredrick, 1967).

The quotations of John Hunter are taken from *Essays and Observations on Natural History, Anatomy, Physiology, Psychology, and Geology (Being His Posthumous Papers on Those Subjects)*, 2 vols. (London: John Van Voorst, Paternoster Row, 1861), and *The Natural History of the Human Teeth and A Practical Treatise on the Diseases of the Teeth*, a special edition, privately printed, for the members of the Classics of Medicine Library (Birmingham, Ala.: Gryphon Editions, Ltd., 1980). Some of the excellent secondary material on the life and work of John Hunter includes: Joseph Adams, *Memoirs of the Life and Doc-*

trines of the Late John Hunter, Esq. (London: J. Callow, 1817); Henry T. Butlin, *The Objects of Hunter's Life and the Manner in Which He Accomplished Them* (London: Adlard and Son, 1907); Roodhouse S. Gloyne, *John Hunter* (Edinburgh: E.S. Livingstone, Ltd., 1950); Sir William Jardine, *Memoir of John Hunter in the Naturalist's Library*, vol. 9 (London: W.H. Lizars, 1843); Jane Oppenheimer, *New Aspects of John and William Hunter* (London: William Heinemann, Ltd., 1946); also see Oppenheimer's *Essays in the History of Embryology and Biology* (Cambridge, Mass., and London: MIT Press, 1967) ; G. Grey Turner, *The Hunterian Museum, Yesterday, Today and Tomorrow* (London: Cassell and Co., Ltd., 1946); Jessie Dobson, "John Hunter's Museum," in Zachary Cope's *The Royal College of Surgeons of England: A History* (London: Blond, 1959). For a greater understanding of the politics surrounding John Hunter's collection, Richard Owen, and the Royal College of Surgeons, see Adrian Desmond, *The Politics of Evolution* (Chicago and London: University of Chicago Press, 1989).

For more information on Arnold Berthold's rooster testicle transplants and the development of endocrinology, see "Hans Selye, Hormones and Stress," in Joel B. Hagen, Douglas Allchin, and Fred Singer, *Doing Biology* (New York: HarperCollins, 1996).

For more information on wet and dry preservation techniques, see *The Preservation of Natural History Specimens*, edited and compiled by Reginald Wagstaffe and J. Havelock Fidler (London: H.F. and G. Witherby). Robert Boyle's explanation of the wet preparation technique can be found in "A Way of Preserving Birds Taken out of the Egg, and Other Small Fetuses; Communicated by Mr. Boyle," in *Philosophical Transactions of the Royal Society* 12, Monday, May 7, 1666. To understand Boyle's contribution to the emerging corpuscular view, see his *The Origin of Forms and Qualities According to the Corpuscular Philosophy* in *The Works of the Honourable Robert Boyle* (London: Johnston, 1772).

CHAPTER 3

The quote by Thomas Malthus is taken from Charles Richard Weld, *A History of the Royal Society*, vol. II (New York: Arno Press, 1975).

Jorge Luis Borges's humorous classification system can be found in "The Analytical Language of John Wilkins," in *Other Inquisitions*, translated by Ruth L.C. Simms (New York: Washington Square Press, Inc., 1966). Michel Foucault's analysis of the Borges quote can be found in *The Order of Things*, translation of *Les mots et les choses* (New York: Vintage Books, 1970 [1966]).

A fascinating discussion of the psychological and philosophical aspects of classification can be found in George Lakoff, *Women, Fire, and Dangerous Things: What Categories Reveal About the Mind* (Chicago and London: University of Chicago Press, 1987). Also see Emile Durkheim and Marcel Mauss, *Primitive Classification*, translated by Rodney Needham (Chicago: University of Chicago Press, 1963).

For further discussion of the political implications of racial classification, see Robert W. Rydell, *All the World's a Fair: Visions of Empire at the American International Expositions 1876–1916* (Chicago and London: University of Chicago Press). Also see Stephen Asma, "Metaphors of Race: Theoretical Presuppositions Behind Racism," in *American Philosophical Quarterly* 32, 1 (1995): 13–30; and John S. Haller Jr., *Outcasts from Evolution: Scientific Attitudes of Racial Inferiority, 1859–1900* (Carbondale: Southern Illinois University Press, 1995). To better understand Louis Agassiz's creationist thinking, see "Contemplations of God," in *The Christian Examiner*, L, 4 (January 1851); and "Evolution and Permanence of Type," *The Atlantic Monthly* XXXIII (1874). The quotation of Louis Agassiz's hopes for a Harvard museum that might rival Hunter's and Cuvier's is taken from Edward Lurie, *Louis Agassiz: A Life in Science* (Baltimore and London: Johns Hopkins University Press, 1988).

An eighteenth-century version of Cesare Ripa's *Iconologia* can be found in the 1758–60 Hertel edition, *Baroque and Rococo Pictorial Imagery*, translated by Edward Maser (New York: Dover Publications, 1971). Konrad Gesner's *Historiae Animalium* can be found in Edward Topsell's *The History of Four-Footed Beasts and Serpents and Insect*, vols. I–III. (New York: De Capo Press, 1967). Ulisse Aldrovandi, *Aldrovandi on Chickens: The Ornithology of Ulisse Aldrovandi (1600), Volume II, Book XIV*, translated by L.R. Lind (Norman: University of Oklahoma Press, 1963). The material on Pieter Paaw's anatomy theater display is taken from William Schupbach, "Cabinets of Curiosities in Academic Institutions," in *The Origins of Museums*, edited by Oliver Impey and Arthur MacGregor (Oxford: Clarendon Press, 1985).

I am indebted to Jacques Roger's masterful book *Buffon: A Life in Natural History*, translated by Sarah Lucille Bonnefoi (Ithaca, N.Y.: Cornell University Press, 1997). The quote about Louis XIII is taken from *Buffon*. And the closing discussion of René de Réamur, Noel Pluche, and the use of the "engine room" metaphor are also indebted to Jacques Roger's *Buffon*.

Further discussion of Dürer's rhinoceros print and its descendants can be found in Brian Ford, *Images of Science: A History of Scientific Illustration* (New York: Oxford University Press, 1993); and T.H. Clarke's expansive *The Rhinoceros: From Dürer to Stubbs, 1515–1799*

(London and New York: Sotheby's Publications, 1986). For more information on the mythical bonnacon and the fanciful natural history of the medieval bestiaries, see Janetta Rebold Benton, *The Medieval Menagerie: Animals in the Art of the Middle Ages* (New York, London, and Paris: Abbeville Press, 1992).

John Ray's *The Wisdom of God* (London: W. Innys, 1743) gives one a good sense of the theological aspects of European natural history during this era. And for a clearer understanding of Ray's classificational theory see his *Methodus plantarum nova* (London: Faithmore and Kersey, 1682; reprinted, Weinheim: J. Cramer, 1962). Much detail and clarity on these issues can be derived from Phillip R. Sloan, "John Locke, John Ray, and the Problem of the Natural System," *Journal of the History of Biology* 5, 1 (1972). Further discussion of the relationship between the early Royal Society of London and the project of a universal classification can be found in Michael Hunter, "The Cabinet Institutionalized: The Royal Society's 'Repository' and Its Background," in *The Origins of Museums*, edited by Oliver Impey and Arthur MacGregor (Oxford: Clarendon Press, 1985).

CHAPTER 4

For more discussion of the influence of Cuvier's cabinet on Emerson, see Lee Rust Brown, "The Emerson Museum," *Representations* 40 (1992).

One can obtain a good understanding of the birth and development of the Royal Society of London from Charles Richard Weld, *A History of the Royal Society*, vols. I, II (New York: Arno Press, 1975).

A thorough study of the logic of Linnaeus's system can be found in Marc Ereshefsky, "The Evolution of the Linnaean Hierarchy," *Biology and Philosophy* 12, 4 (1997). Carl Linnaeus, *Systema naturae per regna tria naturae, secundum classes, ordines, genera, species, cum characteribus, differntiis, synonymis, locis* (Holmiae: Laurentii Salvii, 1758). Passages from Buffon are taken from *Histoire naturelle, générale et particulière* (1788–1804), excerpts in J. Lyon, "The 'Initial Discourse' to Buffon's Histoire naturelle," *Journal of the History of Biology*, 9, 1 (1976). For a better understanding of Lamarck's work, see his *Histoire naturelle des animaux sans vertèbres*, 7 vols. (Paris: Deterville, 1815–1822), and his *Zoological Philosophy*, translated by Hugh Eliot (London: Macmillan, 1914).

My understanding of the original floor plans and organizational plans of the Hunterian and Cuvierian cabinets is indebted to Phillip R. Sloan's excellent essay "Le Muséum de Paris vient à Londres," in *Le Muséum au premier siècle de son histoire*, edited by R. Chartier and C. Blanckaert (Paris: Editions du Muséum national d'histoire naturelle, 1997). Further information on the nineteenth-century layouts of the museums was gleaned from Richard Owen, "Report to the Board of Curators of the Royal College of Surgeons on the Museum d'Anatomie Comparée in the Garden of Plants, Paris," in *RCS Archives*, 275.h.7.(3), fol. 9.; Richard Owen and William Clift, *Descriptive and Illustrated Catalogue of the Physiological Series of the Hunterian Collection*, 6 vols. (London: Taylor, 1833–1840).

I have quoted Cuvier from the following primary sources, which can be found collected in Martin Rudwick, editor and translator, *Georges Cuvier, Fossil Bones, and Geological Catastrophes: New Translations and Interpretations of the Primary Texts* (Chicago: University of Chicago Press, 1997): "Extract from a memoir on an animal of which the bones are found in the plaster stone around Paris, and which appears no longer to exist alive today"; "Note on the Skeleton of a very large species of quadruped, hitherto unknown, found in Paraguay and deposited in the Cabinet of Natural History in Madrid"; and the "Preliminary Discourse." Also see Cuvier's *Leçons d'anatomie comparée*, 5 vols. (Paris: Baudouin, 1800–1805), and his book *The Animal Kingdom Arranged in Conformity with Its Organization*, a translation of *Le Règne animal* (London: G.B. Whittaker, 1827)

The information on unequal conjoined twins comes from Clark Edward Corliss, *Patten's Human Embryology: Elements of Clinical Development* (New York: McGraw-Hill Book Company, 1979). The historical cases of internal parasite twins were taken from G. Gould and W. Pyle, *Medical Curiosities: Adapted from Anomalies and Curiosities of Medicine* (London: Hammond Publishing, Ltd., 1982 [1896]). The strange case of Thomas Lane is taken from the *Catalogue of the Contents of the Museum of the Royal College of Surgeons in London* (London: Richard Taylor, 1830).

Some of the classic expressions of biology-based natural theology can be found in Charles Bell, *The Hand, Its Mechanism and Vital Endowments as Evincing Design: Bridgewater Treatises on the Power, Wisdom, and Goodness of God, as Manifested in the Creation, Treatise IV* (London: Pickering, 1833); and William Paley, *Natural Theology, or Evidences of the Existence and Attributes of the Deity Collected from the Appearances of Nature*, 2nd edition (Oxford: J. Vincent, 1828). The letter from David Hume to his brother, describing John Hunter's examination and diagnosis of cancer, is quoted from Jane Oppenheimer's *New Aspects of John and William Hunter* (New York: Henry Schuman, 1946).

CHAPTER 5

The quotation of Charles Willson Peale, regarding the pedagogical importance of specimen arrangement, is taken from Peale's 1800 letter to a U.S. senator, quoted in Edward P. Alexander's *Museum Masters* (Nashville, Tenn.: American Association for State and Local History, 1983).

Tim M. Berra's handy little primer, *Evolution and the Myth of Creationism* (Stanford, Calif.: Stanford University Press, 1990), provided the statistical information on the poor condition of American biology education, and it also proved helpful in summarizing the crucial debate points between creationism and evolutionary science. For more discussion of the demarcation problem between science and pseudo-science (especially creationism), see Michael Ruse, "Creation-Science Is Not Science"; Larry Laudan, "Commentary: Science at the Bar—Causes for Concern"; and Michael Ruse, "Response to the Commentary: *Pro Judice*," all in *Philosophy of Science*, edited by Martin Curd and J.A. Cover (New York and London: W.W. Norton and Company, 1998). My characterization of the demarcation problem, and the explication of "criteria of adequacy," is indebted to Theodore Schick and Lewis Vaughn, *How to Think About Weird Things* (Mountain View, Calif., London, and Toronto: Mayfield Publishing Company, 1998). Also see Michael Ruse, *The Darwinian Paradigm* (London and New York: Routledge, 1993). The quotations by Darwin are taken from *On the Origin of Species: A Facsimile of the First Edition* (Cambridge, Mass.: Harvard University Press, 1964).

The material regarding the natural selection of human skin coloration is derived from the following sources: A. Roberto Frisancho, *Human Adaptation: A Functional Interpretation* (Ann Arbor: University of Michigan Press, 1986); R.M. Neer, "The Evolutionary Significance of Vitamin D, Skin Pigment, and Ultraviolet Light," *American Journal of Physical Anthropology* 43 (1975); W.F. Loomis, "Skin Pigment Regulation of Vitamin D Biosynthesis in Man," *Science* 157 (1967).

For more information on Richard Owen's concept of "unity of type," see his *On the Nature of Limbs*. (London: Van Voorst, 1849). Also see N. Rupke, "Richard Owen's Vertebrate Archetype," *Isis* 84 (1993), and Rupke's *Richard Owen: Victorian Naturalist* (New Haven and London: Yale University Press, 1994).

Further information on the design and development of the AMNH's Hall of Vertebrate Origins can be found in Henry S.F. Cooper Jr., "Origins: The Backbone of Evolution," *Natural History* 105, 6 (1996).

For a detailed examination of the systematics debates (pheneticists versus cladists, etc.), see *Conceptual Issues in Evolutionary Biology*, 2nd edition, edited by Elliott Sober (Cambridge, Mass., and London: MIT Press, 1994). Especially see Robert Sokal, "The Continuing Search for Order"; Willi Hennig, "Phylogenetic Systematics"; Ernst Mayr, "Biological Classification: Toward a Synthesis of Opposing Methodologies"; and David Hull, "Contemporary Systematic Philosophies." Also see Ernst Mayr, *Toward a New Philosophy of Biology* (Cambridge, Mass., and London: Harvard University Press, 1988) and his *Principles of Systematic Zoology* (New York: McGraw-Hill Book Company, 1969). The quotation by Leonard B. Radinsky comes from his book *The Evolution of Vertebrate Design* (Chicago and London: University of Chicago Press, 1987). For more discussion of the difference between logical classification and genealogical systematics (and the metaphysics of classes versus populations), see Michael T. Ghiselin's excellent book *Metaphysics and the Origin of Species* (Albany: State University of New York, 1997); the Ghiselin quote is taken from this text. For a reasoned relativist perspective on classification systems see John Dupré, *The Disorder of Things: Metaphysical Foundations of the Disunity of Science* (Cambridge, Mass., and London: Harvard University Press, 1993). Some of my discussion of the relationship between observation and theory has been influenced by Harold I. Brown's *Observation and Objectivity* (New York and Oxford: Oxford University Press, 1987).

For more information on the design and development of the AMNH's Hall of Biodiversity, see Brian Morrissey, "The Making of a Rainforest," *Natural History*, June 1998.

The discussion of Darwin's influence, or lack of it, in France draws upon Robert E. Stebbins, "France," in *The Comparative Reception of Darwinism*, edited by Thomas F. Glick (Chicago and London: University of Chicago Press), and Peter Bowler, *The Eclipse of Darwinism* (Baltimore: Johns Hopkins University Press, 1983); the quote from Thomas Hunt Morgan is taken from Bowler, and the quote from Royer is from Stebbins. For a further understanding of nineteenth-century French positivism in biology, see Claude Bernard, *An Introduction to the Study of Experimental Medicine*, translated by Henry Copley Greene (New York: Dover Publications, Inc., 1957).

CHAPTER 6

The Gallup poll information regarding American creationism is taken from G.H. Gallup and F. Newport, "Belief in Paranormal Phenomena Among Adult Americans," *Skeptical Inquirer* 15, 2 (1991).

Stephen Jay Gould's original article "Nonoverlapping Magisteria" appeared in *Natural History* 106, 2 (1997). Gould's thesis has been further developed in his book *Rock of Ages: Science and Religion in the*

Fullness of Life (New York: Ballantine Books, 1999). A fascinating compilation of formal declarations regarding evolution theory by various religious organizations can be found in *Voices for Evolution*, edited by Molleen Matsumara (Berkeley, Calif.: National Center for Science Education, Inc., 1995). For an analysis of the philosophical underpinnings of Darwin's revolution, see Stephen T. Asma, *Following Form and Function: A Philosophical Archaeology of Life Science* (Evanston, Ill.: Northwestern University Press, 1996).

The Einstein quote is taken from A. Einstein, "Science and Religion," in *Focus on Judaism, Science and Technology*, edited by J. Bemporad, A. Segal, J.D. Spiro, and R.S. Wisdom (New York: Union of American Hebrew Congregations, 1970).

For an excellent survey of the literary expressions of Western fears about reductionism, see Roslynn D. Haynes, *From Faust to Strangelove: Representations of Scientists in Western Literature* (Baltimore and London: Johns Hopkins University Press, 1994).

The quote from Richard Dawkins is taken from his *River out of Eden: A Darwinian View of Life*, Science Masters Series (New York: Basic Books, 1995).

My thesis that metaphysical and moral commitments can and should have significant grounding in the empirical world is influenced by John Dewey's "The Construction of the Good," in *The Quest for Certainty* (New York: Minton, Balch, and Company, 1929). The basic position, that moral judgments and judgments of natural fact are different not in kind but only in scope, is generally referred to as ethical naturalism. There is a very distinguished tradition of ethical naturalism that arguably includes the consequentialist philosophers such as J.S. Mill and Jeremy Bentham as well as virtue-based thinkers like Aristotle.

The use of the Wittgensteinian picture metaphor, when discussing different religious and secular worldviews, is influenced by Ludwig Wittgenstein, *Lectures and Conversations on Aesthetics, Psychology and Religious Belief*, edited by Cyril Barrett (Berkeley: University of California Press, 1966); and Hilary Putnam's analysis of Wittgenstein in *Renewing Philosophy* (Cambridge, Mass., and London: Harvard University Press, 1992).

The personal interview with Dr. Eric Gyllenhaal was supplemented by the in-house summative evaluation document *Evaluation of Visitors' Use and Reactions to Life over Time (A New Permanent Exhibition at the Field Museum, Chicago, IL)*, prepared by People, Places and Design Research (Jeff Hayward and Jolene Hart, with the assistance of Eric Gyllenhaal), May 1996.

CHAPTER 7

A good discussion of ancient and medieval memory techniques, using imagery, can be found in M.T. Clanchy, *From Memory to Written Record* (London: Edward Arnold, 1979); also see Frances A. Yates, *The Art of Memory* (Chicago: University of Chicago Press, 1974). For further information on visual imagery and cognition during the Enlightenment, and the general distinction between visual and text-based culture, see Barbara Stafford, *Good Looking: Essays on the Virtue of Images* (Cambridge, Mass., and London: MIT Press, 1996), and *Artful Science: Enlightenment, Entertainment and the Eclipse of the Visual* (Cambridge, Mass., and London: MIT Press, 1994). Parts of my discussion of visual versus linguistic thinking are drawn from David Topper, "Towards an Epistemology of Scientific Illustration," in *Picturing Knowledge: Historical and Philosophical Problems Concerning the Use of Art in Science*, edited by Brian S. Baigrie (Toronto, Buffalo, and London: University of Toronto Press, 1996). The quotation by Neil Campbell and colleagues is from Neil A. Campbell, Jane B. Reece, and Lawrence G. Mitchell, *Biology*, 5th edition (Menlo Park, Calif.: Benjamin Cummings, 1999).

The quotation of curator Sally Love is taken from her "Curators as Agents of Change," in *Exhibiting Dilemmas: Issues of Representation at the Smithsonian*, edited by Amy Henderson and Adrienne L. Kaeppler (Washington, D.C., and London: Smithsonian Institution Press, 1997).

For a general overview of the role of visual illustration in the anatomical and medical sciences, see Ludwig Choulant, *History and Bibliography of Anatomical Illustration*, translated by M. Frank (Chicago: University of Chicago Press, 1920), and Thomas S. Jones, "The Evolution of Medical Illustration," in *Essays in the History of Medicine* (Chicago: Davis Lecture Committee, 1965). Also see Martin Kemp, "Medicine in View: Art and Visual Representation," in *Western Medicine: An Illustrated History*, edited by Irvine Loudon (Oxford and London: Oxford University Press, 1997).

A full discussion of William Hunter's "muscle man" statue can be found in *Dr. William Hunter at the Royal Academy of Arts*, edited and introduced by Martin Kemp (University of Glasgow, 1975).

I highly recommend John and Andrew Van Rymsdyk, *Museum Britannicum* (London, 1778). A good biography of Jan Van Rymsdyk's rather sketchy life can be found in John L. Thorton, *Jan Van Rymsdyk: Medical Artist of the Eighteenth Century* (Cambridge and New York: Oleander Press, 1982).

A developed discussion of the rhetorical use of images in hominid anthropology can be found in Stephanie Moser, "Visual Representation in Archaeology: Depicting the Missing-Link in Human Origins," in *Picturing Knowledge: Historical and Philosophical Problems Concerning the Use of Art in Science*, edited by Brian S. Baigrie (Toronto, Buffalo, and London: University of Toronto Press, 1996).

For more information on Richard Owen's rhetorical construction of dinosaur exhibits, see Adrian Desmond, "Designing the Dinosaur: Richard Owen's Response to Robert Edmont Grant," *Isis* 70 (1979). For an excellent discussion of the persuasive power of imagery in nineteenth-century natural history, see Martin J.S. Rudwick, *Scenes from Deep Time: Early Pictorial Representations of the Prehistoric World* (Chicago and London: University of Chicago Press, 1992).

The quote from Daniel Dennett is taken from his "Intuition Pumps," in *The Third Culture*, edited by John Brockman (New York: Touchstone, 1996). The quote from scientific illustrator Phyllis Wood is taken from her *Scientific Illustration* (New York: Van Nostrand Reinhold, 1994).

Parts of my discussion of affective learning in museums is indebted to Lisa Roberts, "The Elusive Qualities of 'Affect,'" in *What Research Says About Learning in Science Museums*, edited by Beverly Serrell (Washington, D.C.: Association of Science-Technology Centers, 1990). The information from Minda Borun of the Franklin Institute Science Museum is taken from "Naive Notions and the Design of Science Museum Exhibits," in *What Research Says About Learning in Science Museums*, edited by Beverly Serrell (Washington, D.C.: Association of Science-Technology Centers, 1990). Also see Jordan Brealow, "We Don't Want No Educainment!" *The Journal of the Learning Sciences*, August 1995.

Some of the information regarding the Field Museum's exhibit development for "Sue," the *T. rex.*, specifically the quotes of Rich Faron, are taken from William Mullen, "This Fossil Rocks," in *Chicago Tribune Magazine*, April 26, 1998.

INDEX